Concrete and Masonry Problem Clinic

Concrete and Masonry Problem Clinic

The Aberdeen Group

426 South Westgate, Addison, Illinois 60101 Telephone: 708-543-0870 Fax: 708.543.3112

Cover photography: Ab-Train (right photo)

ISBN 0-924659-39-4

10 9 8 7 6 5 4 3 2 1

The Aberdeen Group, *Concrete Construction*® magazine, and *The Magazine of Masonry Construction,*™ and their employees and agents are not engaged in the business of providing engineering or architectural services, nor are they licensed to do so. The information in this book is intended for the use of craftspeople, contractors, and engineers competent to evaluate its applicability to their situation, and who will accept responsibility for the application of the information. The Aberdeen Group disclaims any and all responsibility for the application of the information.

Printed in the USA

Contents

How to order additional copies

Additional copies of the *Concrete and Masonry Problem Clinic* are available through The Aberdeen Group. Call toll free 800-323-3550, extension 119. (In Illinois, Alaska, or Hawaii, call: 708-543-0870, extension 119.) Visa, MasterCard, and American Express are accepted.

Introduction

The Problem Clinic in *Concrete Construction*® magazine and the Troubleshooting Masonry department in *The Magazine of Masonry Construction*™ answer readers' questions about concrete and masonry design, construction, maintenance, and repair. The answers come from technical literature, from editors' experience, and from experts throughout the concrete and masonry industries.

This collection of questions and answers is taken from issues of *Concrete Construction* from 1980 to 1989 and from *The Magazine of Masonry Construction* from 1988 to 1989. The information is grouped in the following categories:

Concrete
- Materials and Mix Design
- Formwork
- Mixing, Placing, Compacting
- Finishing, Jointing, Curing
- Paving
- Troubleshooting
- Repair and Maintenance
- Testing

Masonry
- Materials
- Construction and Repair
- Design

Where appropriate, the original response is altered to reflect dates for the latest documents from the information sources cited.

Acknowledgements

We would like to thank the following people for their help in producing *Concrete and Masonry Problem Clinic*.

Ward R. Malisch, editorial director, *Concrete Construction*® magazine

Katherine Anderson, associate editor, *Concrete Construction*® magazine

Mark A. Wallace, editor, *The Magazine of Masonry Construction*™

Norbert V. Krogstad, senior engineer, Wiss, Janney, Elstner Associates, Inc., Northbrook, Illinois

Part I: Concrete

Materials and Mix Design

1

Concrete

How accelerators protect against freezing

Q. *I've been having an argument. My friend says calcium chloride is added to concrete as an antifreeze. I say it's added to help the concrete gain strength faster so it can have a better chance to survive if it freezes the first night. Am I right?*

A. You're closer to the truth than your friend. It's true that calcium chloride depresses the freezing point of water, but the amount of calcium chloride that can be used to accelerate concrete will not depress the freezing point enough to notice. Hence, that's not the reason for adding calcium chloride.

It's often said that the purpose of adding calcium chloride in cold weather is to help concrete achieve a strength of 500 psi before it freezes, and this is a good rule of thumb. However, 500 psi compressive strength represents a very small tensile strength, one that would be inadequate to resist the large internal forces generated when water freezes. Hence, adequate strength is *related* to the reason for adding calcium chloride but it's not the real reason.

When portland cement hydrates, some of the mixing water gets used up by combining chemically with portland cement. By the time the concrete has achieved a strength of around 500 psi, enough mixing water has combined with the portland cement so that the mixing water fills only 90 percent or less of the pores. That amount leaves enough space available for the water to expand in freezing without damaging the concrete. Hence, the calcium chloride accelerates the reduction in water in the cement paste to a level where an early freeze won't completely disrupt the concrete. The same general principle applies to other accelerators used for cold weather concreting.

Liquid or solid calicium chloride

Q. *What is the difference between liquid and solid calcium chloride accelerators for concrete?*

A. Calcium chloride is a crystalline solid that may be either anhydrous (water free) or dihydrate (combined with a small amount of water). It is available in flake and pellet form. So-called liquid calcium chloride is really a solution made by dissolving either anhydrous calcium chloride or any calcium chloride hydrate in water. The solution usually contains somewhere between 28 and 42 percent anhydrous calcium chloride by weight of solution, depending on the specifications of the buyer. If flakes or pellets are to be used as an accelerator they should be dissolved in water before being added to the concrete, so sometimes it's more convenient to buy a solution. There is a danger if dry calcium chloride is added to concrete because it may not get thoroughly mixed into the batch. This could cause nonuniform setting or strength gain, and concrete cracking or localized corrosion of reinforcing steel if conditions promote corrosion.

Flake calcium chloride added directly to concrete

Q. *What is wrong with putting flake calcium chloride in the concrete truck at the jobsite to offset cold weather effects on setting and strength gain?*

A. Calcium chloride may not dissolve completely if it's added in dry form, either at the plant or at the jobsite. Also, it may not get mixed uniformly throughout the concrete. Nonuniform mixing can make the concrete set unevenly, causing finishing problems. Undissolved lumps can cause popouts or dark spots in the hardened concrete.

Adding calcium chloride in a water solution prevents these problems. To make the solution, use 1 pound of calcium chloride per quart of water and be sure to put the dry material in the water rather than pouring water on the dry material. If you add water to the calcium chloride, a coating may form and keep the material from dissolving completely.

Replacing lost air

Q. *If entrained air is lost due to mixing the concrete too long, can you just add more air-entraining admixture at the jobsite?*

A. Air-entraining admixtures added at the jobsite may not get thoroughly mixed into the concrete. It would also be difficult to control the amount of air entrained under these circumstances. Because of this, it isn't usually a good practice to permit air-entraining admixtures to be added at the jobsite.

Latex-modified concrete

Q. *Is an air-entraining admixture used in latex-modified concrete? If not, why not?*

A. Entrained air isn't required for freeze-thaw durability of latex-modified concrete. The water-cement ratio (including water in the latex) is usually less than 0.40 by weight and this means that there are fewer water-filled voids that are a source of destructive pressure during freezing. The water-filled voids are partially emptied as mix water combines with cement during hydration and, because latex-modified concrete is less permeable, the voids aren't as likely to get filled again with water at a later time.

Effect of vibrating screed on entrained air

Q. *Doesn't vibrating screed cause a loss of entrained air at the surface, right where you need it most?*

A. When vibration is properly applied it removes mainly large air bubbles that don't contribute to freeze-thaw durability of the surface. The air-void-spacing factor, one of the most critical characteristics of air-entrained concrete, remains relatively constant even though the average size of air voids gets smaller. Prolonged vibration may harm the air void system. To avoid this, don't leave vibrators running when the screed isn't moving forward and don't make multiple passes over the same concrete with the screed.

Air loss in fly ash concrete

Q. *We've had trouble with low entrained air contents in concrete containing fly ash. What causes this problem and what can we do about it?*

A. Unburned carbon in fly ash will absorb air entraining agents and may necessitate an increase in the air entraining agent (AEA) dosage. Loss on ignition (LOI) is a measurement of unburned carbon remaining in the ash. Specifications of the American Society for Testing and Materials (ASTM C 618) for fly ash used in concrete limit the LOI to 6 percent. Values greater than 2 to 3 percent, however, will probably require you to significantly increase the air entraining agent dosage. You might also try changing the type of AEA used. Several admixture manufacturers have developed AEA's specifically for use in fly ash concrete.

Durability of dry-mix shotcrete

Q. *Is it possible to entrain air in shotcrete when the dry-mix method is used? In 30 years experience I have never learned of a way to do it. I am interested in anything that can enhance the durability of shotcrete.*

A. We do not believe it is possible. Fortunately, it doesn't seem to be necessary, as indicated by a Portland Cement Association laboratory study of the durability of shotcretes to freezing and thawing. They found that the dry-mix shotcretes studied had excellent durability, even though the air contents were very low. The small amount of air that these mixes did contain was found to be *entrapped* (not entrained), so the air could not have contributed to the durability. Details are given in "Laboratory Study of Shotcrete," *Symposium on Shotcreting, Publication SP-14*, American Concrete Institute,

1966. This has also been reprinted as Portland Cement Association *Development Department Bulletin D111.*

Air-entrained high strength concrete

Q. *We are interested in the possibility of using high strength concrete in pavements and bridge decks to improve durability. Has anybody achieved 10,000-psi concrete with an air content of 8 percent? If not, is it possible that it can be done?*

A. Several people knowledgeable about high strength concrete feel that the strength of 10,000-psi concrete would suffer drastically at such air contents. However, two of our readers have produced concrete to similar specifications.

Prairie Material Sales, Inc. was asked by the Portland Cement Association to produce concrete with a unit weight of 145 pounds per cubic foot, 7½-inch slump, and 10,000 psi strength. The mix design had to incorporate 6 ± 1 percent air and the superplasticizer had to achieve the 7½-inch slump. This combination assured both a bond around the heavy reinforcement and the required pumpability of the mix. The concrete performed in all phases of the specification requirements. It should be noted that the undertaking by this company was the result of many years of product development and not all admixtures or raw materials will perform with the success they had in this project.

Another company, Hardy Associates Ltd., has successfully produced mixes with 7 ± 1 percent air for use in major concrete structures. They achieved compressive strengths of approximately 8700 psi at 28 days and 9700 psi at 56 days. Two common features of their mixes are the fly ash and the use of superplasticizing admixtures. The 28-and 56-day strengths were achieved with water-to-cementitious-material ratios of approximately 0.29. Normal materials were used to achieve the performance levels: Type 1 portland cement, concrete sand, and partly crushed coarse aggregate. As one would expect, even small variations in air content proved to be quite significant with respect to compressive strength.

Hardy Associates has also conducted research on high strength semi-lightweight concrete with high air content. Typically these mixes with a plastic unit weight of less than 125 pounds per cubic foot had air contents of 6 ± 0.5 percent and achieved compressive strengths of approximately 6500 psi at 28 days. Various lightweight coarse aggregates were used in combination with normal density sand. It should be emphasized that this was project-orientated contract research.

This company found that compressive strength *per se* was not a problem. By incorporating superplasticizers and silica fume the mixes were produced with compressive strengths in excess of 22,000 psi at 3 days. Difficulties can arise when one attempts to achieve high air contents in combination with very high strengths and/or very low slumps. However, such difficulties may be overcome using today's technologies.

Proportioning mixes with Class C fly ash

Q. *A new source of Class C fly ash is available. Would we use the same amount as the Class F that we have been using?*

A. It would be best to make some trial mixes. In general, the weight of Class F fly ash is usually more than the weight of cement replaced, whereas the weight of Class C fly ash used is usually less than the weight of cement replaced. Thus you'll probably need less Class C than the amount of Class F you've been using.

Fly ash in shotcrete and bond of finish coat

Q. *A swimming pool contractor who is a customer of ours wants an oversanded mix to use in shotcreting. We have suggested one of our oversanded mixes which contains 730 pounds of cement, 180 pounds of fly ash, 700 pounds of ½-inch stone, and 1800 pounds of sand. We often supply this mix at a 6-inch slump to other customers who use it with a small concrete pump. This customer is afraid that his finish coat, which contains marble dust, will not bond properly to a shotcrete containing fly ash. Should we reproportion the mix to eliminate fly ash?*

A. We don't see any problem with your mix. You might caution the contractor to keep the shotcrete moist so that the surface cures well and also to keep the surface clean. He could accomplish both by keeping it covered with polyethylene.

Minimizing bleeding in post-tensioning grout

Q. *Is it possible to grout vertical post-tensioning tendons without having bleed water collect at the top? We're concerned that the grout be effective for the full length of the tendons. If there is any significant amount of bleeding, the tops of the tendons could be unprotected and unbonded.*

A. An article on this subject was published in *Concrete Construction* in the January 1984 issue, page 47. Water-retentive grouts of the type described in that article, reported to be essentially free of bleeding, are available.

Recently another grout has been developed which is also reported not to bleed. It is formulated differently, using portland cement, microsilica (also called silica fume) and superplasticizer. Experimental work on it was described by Pierre-Claude Aitcin, G. Bal-livy and R. Parizeau in "The Use of Condensed Silica Fume in Grouts," an 18-page chapter in the book *Innovative Cement Grouting,* Publication SP-83, available from the American Concrete Institute, Box 19150, Detroit, Michigan 48219.

Does silica fume darken concrete?

Q. *Some of the concrete in a project we'll be building will contain silica fume (microsilica) as an admixture. I've been told that the silica-fume concrete will be darker in color than normal concrete even though the same cement is used in both. How much difference in color can I expect?*

A. Freshly mixed concrete containing silica fume can be almost black, dark gray, or practically unchanged, depending on the dosage of silica fume and its trace minerals content. We asked Harry Tormey of W. R. Grace & Company, a manufacturer of silica-fume admixtures, to send a hardened concrete sample showing the worst-case color difference that might be expected (see photo). The dark side of the sample contains silica fume at a dosage of 105 pounds per cubic yard. The sawed surface has been given a clear coating to accentuate the difference. Note that the color difference isn't as pronounced on the uncoated edge of the sample. Tormey says that the greatest differences in color occur in concretes made with cements that are light in color. Mix proportions may also have an effect. If the color difference is a problem, you might want to use the darkest brand of cement available in your area and try different mix proportions during the mix design process.

This sample shows a worst-case color difference between normal (left side) and silica-fume (right side) concrete. A clear coating was used on the sawed cylinder surface to accentuate the color contrast. The edge of the cylinder is uncoated.

Reduce cover with silica-fume concrete?

Q. *Is there any current practice or potential for reducing required concrete cover when silica-fume concrete is used? With its greatly reduced permeability, shouldn't you be able to get by with less clear cover? This would be especially helpful for reducing weight in precast products.*

A. Consultant Terry Holland says there have been no reductions in cover requirements for structural members containing silica-fume concrete and none is expected. Silica-fume concrete hasn't been in service long enough to permit evaluation of its overall impact on long-term concrete performance. Only after a great deal of long-term data is available should any such reduction be considered. Until then, use the belt-and-suspenders approach: Add silica fume to cut permeability and resist chloride intrusion, but specify adequate cover as well.

Microsilica

Q. *What is microsilica—a cement or an admixture? Where can I get more information about it?*

A. "Microsilica" is another name for silica fume. Some people in the concrete industry believe that "microsilica" is a better and more descriptive term than "silica fume." The product comes off in the fumes of electric furnaces that produce ferrosilicon or silicon metal and thereby gets its name. However, by the time it is collected for sale to people such as those in the concrete industry it has become a very finely divided solid—microsilica. Much has been written about the use of silica fume as a pozzolanic admixture in concrete. You can find a good discussion in a 3-page article beginning on page 443 of our May 1982 issue.

Detecting superplasticizer in concrete

Q. *Is there any way to tell, after the concrete has hardened, whether a superplasticizer was added to the concrete? Is there a way to tell how much? How about the water-cement ratio?*

A. Yes, it is possible to determine whether a superplasticizer has been used in concrete, but few laboratories are equipped to do it. The process involves making a solvent extraction and then doing an infrared analysis of the extract to determine whether a melamine or a phenol formaldehyde system is present.

In some cases it is possible to make a quantitative determination. If you already know the specific identities of the chemicals used, it is possible to determine the amount of superplasticizer.

There are two possible methods for determining the water-cement ratio in a sample of hardened concrete. One is a linear traverse method described by Axon in a paper in the ASTM *Proceedings* (1962), pages 1068-1080. The other is a method described by Brown in a Building Research Station (United Kingdom) publication, *Analysis of Concrete*. This latter method involves vacuum saturation of the concrete with carbon tetrachloride, but the method doesn't work on air-entrained concrete because some of the air voids take up carbon tetrachloride and consequently get counted as part of the water.

A good *estimate* of water-cement ratio can be made petrographically by measuring the size of the calcium hydroxide crystals. A petrographer familiar with the method might come up with an answer somewhere between 0.55 and 0.65 when estimating a concrete whose actual water-cement ratio is 0.60.

Flowing concrete

Q. *Please explain flowing concrete. Is it the same as superplasticized concrete? If it flows so well, doesn't it segregate? Is it really as strong as concrete made with a lower slump?*

A. So-called flowing concrete is concrete that is made to a slump that is temporarily higher than normal. It is usually made by adding admixtures after the concrete arrives at the job site at a slump, for example, of 2 inches. The proper amount of admixture is added and thoroughly mixed in. The slump increases to perhaps 6 or 8 inches but then declines over a period of about 1/2 hour to the original 2 inches or less. In the meantime the slump is high enough to greatly facilitate placement operations. Although there have been some reports of segregation, it is said this seldom happens (see "How Super are Superplasticizers," *Concrete Construction,* May 1982, page 409).

Admixtures used for flowing concrete are sometimes called high-range water reducers and sometimes superplasticizers. Flowing concrete can also be made, however, by using two separate conventional admixtures at the same time; these jointly serve to provide the needed slump and control the set for the particular conditions of the job. The choice of two admixtures is said by some to have advantages over high-range water reducers but the particular admixtures used must be selected specifically for the job.

Although flowing concrete should theoretically have about the same strength as normal concrete, reports indicate that it is sometimes a little bit stronger.

Black concrete driveways

Q. *A job calls for 200 cubic yards of black concrete for driveways. We have been told that this would require 10 percent black pigment by weight of cement. If we use lampblack as pigment will it knock out the entrained air? What can we do to be sure the job will be colorfast?*

A. For a black surface made with integral pigment the best material to use would be a black admixture based on black mineral oxide and containing a water-reducing agent to disperse the color thoroughly. Such an admixture will not knock out the entrained air, but you should of course check the trial mixes for proper air content. All pigments tend to become partially obscured on flatwork in time due to a small amount of lime moving to the surface and being carbonated there. The effect can be minimized by applying a black colored wax after the concrete has hardened. This should be available from the color admixture supplier.

Aggregates

Local aggregate unsuitable for fire endurance

Q. *For a job in New Mexico we must site-cast some double-tees. The coarse aggregate in the area contains 61 percent free silica but the Underwriters Laboratories permit no more than 20 percent free silica for fire endurance. Is there any way to treat the aggregate to reduce the silica content? The only alternative is to truck a suitable aggregate in from about 150 miles away.*

A. The only method of beneficiating the local aggregate that we know about is heavy-media separation. However, the constituents of the aggregates in the New Mexico area are reported to be so close together in specific gravity that heavy-media separation is inefficient and expensive. The best solution appears to be to truck in a limestone or other suitable aggregate.

Alkali-silica reaction appears in some new locations

Q. *We have an unusual case of an apparently expanding concrete that baffles local experts. The contractor removed all the old concrete of a deteriorating sidewalk and placed a new one in December 1981. Due to cold weather, the contractor ordered 4000-psi concrete and built a tent around the entire job. Salamanders were placed in the tent but no records were kept of the actual temperatures inside. The contractor did not cover the new sidewalk so the concrete was apparently subject to drying at an unknown temperature in its early stages. The base material was well compacted sand and gravel. The sidewalk, located on a hill, was placed in 16-foot-long sections separated by 1/2-inch joints that were subsequently sealed.*

Later in 1982, it was noticed that the individual sections were apparently enlarging and each had expanded downhill and in many cases the higher section had overridden the lower one, squeezing out the joint sealant. Visually the whole thing is a mess. Since this is a show sidewalk in the center of town, it will be necessary to either replace it or make repairs if we can find a satisfactory method.

We have established conclusively that the concrete is not simply sliding down the hill, but each section is apparently expanding. Since concrete made with quality aggregate is not supposed to expand beyond its original dimensions, we don't know what to make of this situation. Do you have any ideas?

A. The expansion is much larger than could be accounted for by changes in moisture or temperature. We checked with William Hime, of Erlin Hime Associates, Northbrook, Illinois, about whether there has ever been trouble with alkali-silica reactivity with any aggregates in your area. He replied that such reactivity has become more likely than heretofore because of the recent trend toward higher alkali contents of cements.

Most areas of the country have been free of this trouble. But in recent years, new dust-control requirements by the Environmental Protection Agency have required changes in the manufacture of portland cement. These in turn have tended to increase the alkali content of some cements. Now some aggregates that had formerly undergone no noticeable alkali-silica reaction (because they had always been used with low-alkali cements) are beginning to reveal their reactivity when used with higher-alkali cements.

You might check this out by testing the cements and aggregates you use by the mortar bar test of ASTM C 227.

Sparkling aggregate for flatwork

Q. *While visiting downtown San Francisco, we noticed that the concrete sidewalks appeared to have a sparkly substance in them. We really like this effect. Can you help us identify the substance and tell us how it can be made available to us? Is this substance compatible with epoxy poly-pebble systems?*

A. Our investigations indicate that the effect is obtained by using a proprietary product made in San Francisco containing cement and fine-grained silicon carbide (usually $^{16}/_{36}$ grit) with or without color pigment. It is normally supplied in 100-pound bags and added at the rate of 0.20 to 0.25 pound per square foot of flatwork. In addition, a factory-blended color hardener may be included at the rate of about 0.50 pound per square foot though pastel-colored materials such as light greens and buffs require 0.60 to 0.75 pound per square foot to be effective. The manufacturer's instructions call for applying the product as a dust-on shake after leveling and screeding and before bleeding starts. It is lightly worked into the surface by float. On large areas this can be done by a power trowel with float shoes attached. Finish troweling is delayed to avoid burying the sparkling grains. For best resistance to freeze-thaw exposures the manufacturer recommends applying a cure-and-seal agent and repeating this application once a year. Normally air-entrained concrete would be used for such exposures, but the company recommends using the cure-and-seal agent whether the concrete is air-entrained or not.

As to your second question, the material is compatible with epoxy-aggregate toppings but it requires a great deal more material to obtain the effect.

Batching dry aggregates in concrete

Q. *Is it acceptable to batch concrete with aggregates below the saturated surface-dry moisture state? If not, what standard prohibits it?*

A. For normal-weight concrete, batching dry aggregates shouldn't cause any problems. Most air-dry aggregates don't absorb more water than a percent or so by weight of the aggregate. The batch plant operator can correct for the amount of water absorbed by adding more water at the plant.

Batching very dry lightweight aggregates can cause problems if the concrete is to be pumped. Absorption of mix water by the aggregate may cause rapid slump loss and rock jams in the pumpline. But it's difficult to predict exactly how much water will be absorbed and correct the amount of water added at the plant to eliminate the problem. To avoid this problem, presoak lightweight coarse aggregate by sprinkling if the concrete will be pumped.

An even better solution is saturating the aggregate at the plant where it's produced, using either a vacuum or thermal method. Then keep it from drying before it's batched to prevent unnecessary pumping problems. For more information on pumping lightweight concrete, see our June 1985 article, "Pumping Structural Lightweight Concrete," on page 505.

Segregation based on aggregate source

Q. *I have a problem with concrete segregation that appears to be related to the coarse aggregate. When we use crushed stone from one source the concrete segregates in the chute of the mixer truck. When we use stone from another source we get no segregation. Both aggregates meet the ASTM C 33 grading specifications. I would like to find some way to use the stone that causes segregation because it costs less and it is closer to the plant. The concrete is air-entrained and we use the same sand in both mixes.*

A. You do not say anything about mix proportions. Possibly you are using the same ratio of coarse to fine aggregate for both coarse aggregates. Even though both aggregates fall within the ASTM C 33 limits there may be a considerable difference between their two gradings. Very likely the two aggregates also differ somewhat in angularity. Either or both of these would mean that each one has its own best ratio of stone to sand. We suspect that the mix which you say segregates is a harsh mix that contains too little sand.

Increasing the sand content may not turn out to be the only way or the best way to improve the mix. Possibly you could adjust the grading of the problem aggregate so that it more nearly resembles that of the better aggregate. If you could find some third source of aggregate, perhaps an aggregate of an intermediate size, you might be able to add a small amount of it in place of some of the problem aggregate and thereby improve the total grading. Experimenting this way with laboratory size trial mixes might lead to an easy and inexpensive solution.

Pea gravel concrete properties

Q. *I want to start a small-line pumping service in my area of Pennsylvania. Is concrete made with 1/2 inch-minus aggregate (pea gravel) suitable for residential and light commercial work?*

A. Small-sized-aggregate concrete is suitable for residential and light commercial work if it's properly proportioned. Concretes made with small-sized aggregates such as pea gravel require higher sand contents for workability. They also require a higher water content. Concrete with 3/4-inch-maximum-size aggregate might need 300 pounds of mixing water per cubic yard to give a 3- to 4-inch slump. Concrete with 3/8-inch-maximum-size aggregate might require 340 pounds of water to give the same slump. If 500 pounds of cement per cubic yard produced the needed strength for 3/4-inch-aggregate concrete, you'd need about 566 pounds of cement for the 3/8-inch-aggregate concrete.

Concrete made with small aggregate shrinks more because of the higher water and cement content. Thus it's more likely to crack in service.

You'll also need a percent or so more air content to protect against freezing and thawing. For 3/4-inch-aggregate concrete the average entrained air contents should be 6%. For 3/8-inch aggregate the average should be about 7.5%. Talk to your local ready mixed concrete producer. He should be able to provide a pumpable mix that gives the strength and durability needed.

Does 151 pcf require heavyweight aggregate?

Q. *There has been a rash of inquiries in our area about where to purchase heavyweight aggregates. Inquiries have also been made about how to make concrete with a unit weight of 147 pounds per cubic foot (pcf) or 151 pcf. We think some federal agency may be the source of such unusual specifications. We wonder if such exact minimum unit weights are necessary, and if so whether there is any need to use heavyweight aggregate to obtain them. Even the higher unit weight seems almost within the normal range.*

A. Apparently some specifications have been written that call for unusually explicit unit weights. One set of specifications, written for a Department of Energy contract for a power company's experimental flue gas cleanup facility, called for "an in-place concrete density of not less than 147 pcf according to ACI 211.1-81." This applied to all concrete to be used for radiation shielding in various areas that were named in the specifications. A total of only about 200 cubic yards was needed. By using more cement and less sand the concrete producer was able to make concrete with unit weights ranging from 147.5 to 152 pcf without using any imported or special aggregate.

A minimum cured density of 147 pcf was also called for on another construction job, a hospital, in the specifications for the walls and roof of the linear accelerator room and passageways. The concrete supplier decided to design a mix with normal weight aggregate, aiming for a design unit weight of 151 pcf by replacing some of the sand with cement. The actual unit weight was 152 pcf. One sample is reported to have been coated with a sealer, weighed after 1 year, and found to still have a unit weight of 152 pcf.

For significantly higher unit weights heavier aggregates would undoubtedly have to be used, at least in part.

Cement

Type I versus Type II cement

Q. *Does Type II cement have an advantage over Type I cement for concrete used in manure storage areas? Is concrete made with Type II cement more acid resistant?*

A. Type II cement is more resistant to chemical attack by sulfates but not necessarily to acid attack. However, while manure may slowly disintegrate concrete, the attack isn't severe enough to warrant the use of a special cement. No mention of cement type is made in the Portland Cement Association publication *Concrete Liquid-Manure Tanks*, which details recommended concrete mixes for feedlot improvements.

What Type III does for strength and when

Q. *Is there any general guide or rule of thumb about how much more strength you get from Type III than Type I cement at various ages?*

A. At ordinary temperatures of about 70° F you can expect about 190 percent of the Type I strength at 1 day, 150 percent at 3 days, 120 percent at 7 days, and 110 percent at 28 days. At 3 months the strength of Type I and Type III would be about the same. At 40° F the advantage of Type III over Type I is more pronounced and the advantage lasts for a longer time.

Disadvantages of using Type III cement?

Q. *Are there any disadvantages in using Type III cement? If so, what are they?*

A. Type III cement is used when high early strength is desired. It gains strength faster because it contains larger amounts of faster reacting compound and because it is usually ground finer than normal cement. Heat is generated faster when the cement hydrates and this can cause problems in any massive structures such as base mats, large piers, and dams. Thermal cracking is more likely when heat builds up quickly in a structure. Finely ground cements also tend to set faster and this can cause problems during hot weather concreting operations.

Improving sulfate resistance

Q. *We must furnish concrete made with Type II cement having a tricalcium aluminate content of 4 to 5 percent. This is to be used in a floor that must be sulfate resistant. Would it make the floor more sulfate resistant to increase the concrete design strength from 3000 to 4000 psi? Would using fly ash in the Type II cement mix help improve the sulfate resistance still further?*

A. The answer to both questions is yes. The Portland Cement Association's *Design and Control of Concrete Mixtures,* Eleventh Edition, Figure 22, page 32, affirms the first answer. The U.S. Bureau of Reclamation *Concrete Manual,* Eighth Edition, affirms the second answer in its Table 2 on page 11.

Type II cement may sometimes substitute for Type V

Q. *The specs require Type V cement. There is none available in this vicinity. Is there anything I can use instead, so I won't have to ship Type V a long distance?*

A. According to the Portland Cement Association, some Type II cements will serve the same purpose as a Type V. ASTM C 150 makes the following requirements for Type V but not for Type II:

- The SO_3 content must be less than 2.3 percent.
- The C_3A (tricalcium aluminate) content must be less than 5 percent.
- The amount of $2C_3A$-plus-C_4AF must be less than 20 percent.
- The sulfate expansion must be less than 0.045 percent.

The American Association of State Highway and Transportation Officials (AASHTO) also requires a lower limit on heat of hydration for Type V cement, but according to the Portland Cement Association this is usually of no consequence.

If a particular brand of Type II cement will meet the additional requirements for Type V, or can be made to meet them, than Type II cement can be substituted. Buyers can ask, if they wish, for assurance that the Type II does meet the requirement by requiring that it be certified on a mill report. If the cement arrives in bags it will probably be labeled Type II because of packaging routines, but the mill report will supersede the information on the label.

High early strength, sulfate resistant cement

Q. *We are site-casting concrete panels for a sewage sludge digester tank. Normally we have used ASTM Type II cement to resist the sulfates present but we would like to use Type III cement to speed up the casting cycle. Must we also use some sort of admixture to make the concrete as sulfate-resistant as Type II?*

A. You can use fly ash if you test the effectiveness of the specific fly ash to be used in order to know how to design the particular mix. This could perhaps be done by using various cement-fly ash ratios in ASTM C 452, "Standard Test Method for Potential Expansion of Portland Cement Mortars Exposed to Sulfate." The tests should be planned and interpreted by an expert on sulfate resistance.

Another alternative would be to use a special Type III cement. ASTM C 150, "Standard Specification for Portland Cement," does make provisions for some optional chemical requirements; when specifically requested, Type III cement with a maximum tricalcium aluminate content of 8 percent may be requested to obtain moderate sulfate resistance equivalent to that of Type II cement. Probably not all cement plants would be willing to provide such a cement.

Forget fly ash for sulfate resistance?

Frank D. Gaus, President of Superior Products Company, Inc. didn't agree with our answer to the Problem Clinic on "High Early Strength, Sulfate Resistance Cement." He tells us:

The answer to the question on sulfate resistant cement is erroneous with respect to fly ash. Fly ash does not do anything for alkali reactivity, and it does absolutely nothing, which is a little less, for resistance to sulfate attack.

A paper by Katharine Mather, "Current Research in Sulfate Resistance at the Waterways Experiment Station," was published in the *George Verbeck Symposium on Sulfate Resistance of Concrete*, ACI Special Publication SP 77, pages 63 to 74. This showed that of eight fly ashes tested with one cement, none significantly changed the sulfate resistance. With another cement, some fly ashes decreased sulfate resistance; and with a third cement six did nothing and two *may* have brought sulfate resistance within tolerable limits.

A suitable natural volcanic ash, calcined, and ground just like the manufacture of portland cement, will do many things in concrete. Mostly, it will stop alkali reactivity, and I mean *stop* it. It will definately increase resistance to sulfate attack. As far as getting early strength is concerned, using calcined natural pozzolan, a mix can be designed with appropriate superplasticizer, water reducing agents or nonchloride accelerators to give high early strengths comparable to high early cement.

Type V portland cement does not substantially help resistance to sulfates—it is like putting Mercurochrome on a 6-inch bullet hole in your chest.

Editor's note: We still believe, after examination of many references, that some fly ash and cement combinations can have improved resistance to sulfates. Since fly ash composition varies significantly depending on its source, an experience record or test program may be needed to establish the sulfate resistance of a particular combination of cement and fly ash.

More on sulfate resistance

A letter regarding the "High Early Strength, Sulfate Resistant Cement" Problem Clinic and the response by Frank Gaus, came from Paul Klieger of the R & D/CTL Group, Portland Cement Association. He tells us:

With respect to the Problem Clinic item, ASTM C 452 is not a suitable test for evaluating the sulfate resistance of blended cementitious systems. With respect to Mr. Gaus' response, he is not correct in three of his statements:

1. The Katharine Mather paper he referred to deals almost exclusively with ASTM C 618 Class C fly ashes (generally high lime fly ashes derived from burning of lignite or subbituminous coals). ASTM C 618 Class F fly ashes, generally derived from the burning of bituminous coal, have CaO contents less than 10 percent. Other studies have also shown that Class C fly ashes do not improve, and indeed may degrade, the sulfate resistance of concrete. Many Class F fly ashes, on the other hand, can be helpful in improving the sulfate resistance of concretes made with nonsulfate-resistant portland cements (C_3A contents greater than 8 percent), so long as adequate curing has been provided to produce the desired level of impermeability to intrusion of sulfates.

2. Fly ashes do function to inhibit the alkali-aggregate reaction. While they are generally not as effective as most calcined shales and clays in this regard, they can be useful in a situation where recourse to the use of an inhibitor appears necessary. Again, Class F fly ashes are probably more effective than Class C, probably due to their higher silica contents.

3. Type V portland cement is designed to provide significantly improved sulfate resistance to concrete. That it does so is demonstrated in Figure 3 of the paper by George Verbeck, "Long-Time Study of Concrete Durability in Sulfate Soils," in ACI Special Publication SP-77, to which Mr. Gaus referred. A copy of this figure is shown here. A rating of "1" is excellent performance. The cement designated "51" is an ASTM Type V cement. Note the importance of both C3A content and the cement content with respect to sulfate resistance. Many other studies have shown similar performance data.

Other high early strength, sulfate resistant cements

Jere H. Rose, Manager of Technical Services for Atlantic Cement Company has written to suggest an additional answer to the "High early strength, sulfate resistant cement" Problem Clinic. He writes:

While I do not disagree with the Problem Clinic answer, you should be aware that there is another material that may be better suited for this application. This is finely ground blast-furnace slag which is readily available throughout the Eastern Seaboard.

The use of ground granulated slag in portland cement concrete has been proven to give high resistance to sulfate attack when used in proportions of 60 percent of total

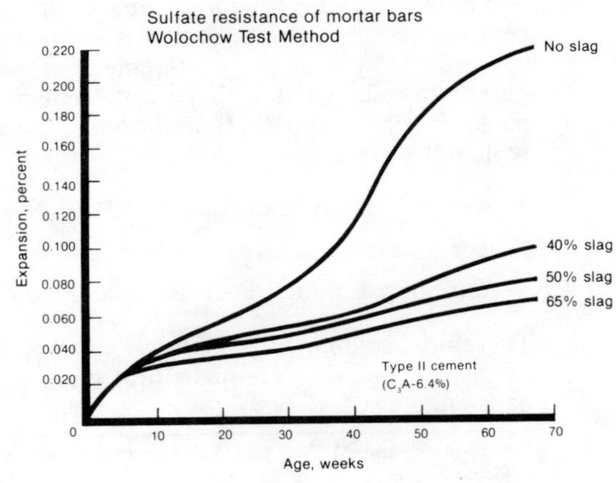

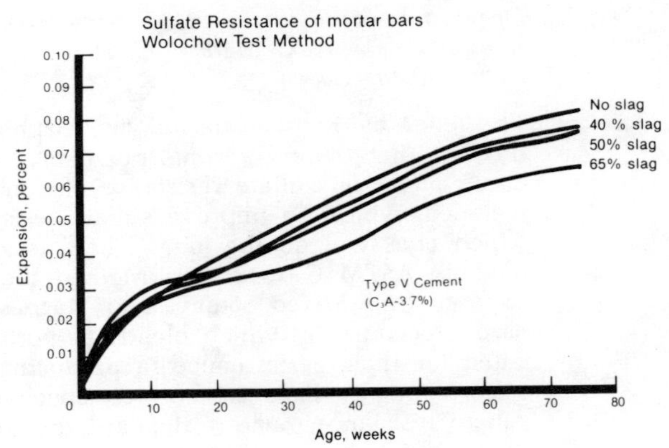

cementitious material. This has been well documented in European practice. Comparison may be made in the two figures shown. If this kind of concrete is kept at a temperature of 72° F or higher, it gains strength faster than similar concrete made with Type I portland cement only. Ground granulated blast-furnace slag also contributes other interesting properties to concrete.

Type V cement and fly ash improve sulfate resistance

Katharine Mather had her own response to the articles appearing on this topic:

Mr. Gaus says in your September 1983 issue that Type V portland cement does not substantially help resistance to sulfates. Mr. Gaus is wrong. In "Guide to Durable Concrete" (ACI 201.2R-77), which I helped write, the recommended protection against severe sulfate attack (1500 to 10,000 parts per million of sulfate in water) is:

- maximum water-cement ratio of 0.45 by weight
- use of Type V cement

For very severe exposure (over 10,000 parts per million of sulfate) the recommendation is for the use of Type V cement plus a pozzolan. The report notes "Best results have been obtained when the pozzolan is a fly ash meeting . . . ASTM C618, Class F.'' This comment is based on U.S. Bureau of Reclamation work reported in 1975-76. My 1982 paper published in ACI Special Publication SP-77 and mentioned by Mr. Gaus, showed that highly sulfate-resistant mortars can be made using high C3A (tricalcium aluminate) Type I cements and 30 percent replacement of cement by a silica fume or a highly siliceous natural pozzolan.

Mr. Gaus also stated that fly ash does not do anything for alkali reactivity. He is wrong about that also. In my paper, which he cites, I remark that effectiveness of pozzolans in preventing excessive expansion due to sulfate attack depends on certain factors and that quite similar factors appeared to be related to pozzolan effectiveness in reducing expansion due to the alkali-silicate reaction. In work reported by L. Pepper and B. Mather* it was shown that typical Class F fly ashes could give a 75 percent reduction in expansion of mortar made with high-alkali cement and Pyrex glass aggregate. On average, replacement of 40 to 44 percent of the cement would be required to achieve the 75 percent reduction in expansion. In the discussion of that paper it was noted that field concrete is seldom subjected to the conditions under which the tests were conducted:

- over 1 percent alkali (as N_a2O) in the cement
- aggregate as reactive as Pyrex glass
- continuous storage at 100° F and 100 percent relative humidity

Therefore, seldom will so large a proportion of the cement need to be replaced with fly ash to achieve adequate protection against excessive alkali-silicate-reaction expansion.

* Pepper, L. and Mather, B. ''Effectiveness of Mineral Admixtures in Preventing Excessive Expansion of Cement Due to Alkali-Aggregate Reaction, '' *Proceedings,* ASTM, V.59, 1959, pages 1178-1203.

New test for sulfate resistance

One last response to the ''High early strength, sulfate resistant cements'' Problem Clinic was received from Bryant Mather, Chief of the Structures Laboratory at the Waterways Experiment Station. He writes:

The letter by Frank Gaus in your September 1983 issue incorrectly calls attention to the fact that there are some pozzolans which have a bad effect and some have little or no effect on sulfate resistance of concrete. Since there are also pozzolans, fly ash included, which do improve sulfate resistance, the question is ''How do you tell which ones will do the job?'' There is a way. It's described in a standard test method, ASTM C 1012, Test Method for Length Change of Hydraulic-Cement Mortars Exposed to a Mixed Sodium and Magnesium Sulfate Solution. This provides the standard procedure by which blends of portland cement with pozzolan or slag can be evaluated for their performance in producing sulfate-resisting cement paste in mortar, concrete, or grout. Expansion of test specimens is measured and used as an indicator of sulfate resistance. Such a standard method of testing has long been needed.

It is now possible for anybody who wants to sort pozzolans that have a good effect in improving sulfate resistance from those that have no effect and from those that have a bad effect. The only drawback is that currently the question of how much expansion is too much hasn't been officially answered. Thus it is up to the people running the test to decide on an acceptable expansion limit and on how long to run the test; my personal suggestion would be 0.1 percent and one year.

Worries about calcium aluminate conversion

Q. *Because it resists corrosion, calcium aluminate cement has been suggested for use in floors in a food processing plant. Some of these floors are above grade. Isn't it true that calcium aluminate loses strength drastically as it gets older? If so, it doesn't seem to be a good choice for a suspended floor.*

A. Hydrated calcium aluminate cements do undergo something called *conversion*—a process by which the hydrates slowly change their crystal structure. Unfortunately, in the process the concrete becomes weaker. Knowledgeable designers overcome this problem by designing accordingly. They call for a mix that will still have the strength needed *after* it has changed its crystalline structure.

Magnesium phosphate cement

Q. *We would like to find a material that sets and hardens rapidly that we can use for repairing concrete in city streets and bridges without disrupting traffic any longer than absolutely necessary. We would prefer that this not be an epoxy or other organic material because we are concerned that differences in thermal expansion and contraction might eventually cause problems.*

A. You could consider a magnesium phosphate cement. Such a cement is used in the same manner as portland cement, though the concrete requires less mixing water, hardens rapidly, requires no moist curing and can be put in service within about an hour at ordinary temperatures. It can also be used for making patches at such temperatures as 10° F when necessary. The cost is probably closer to that of epoxy concrete than that of portland cement concrete.

Hot cement

Q. *Our firm is contemplating a job in Saudi Arabia where the cement would be shipped directly from the mill without storage before shipping. Is use of hot cement likely to cause cracking of the concrete in this hot, dry climate?*

A. We assume that standard good job practices will be followed to minimize drying shrinkage cracking. If that is so, the cement itself should not cause trouble if:

- it meets the physical and chemical specifications for the project
- the cement is at a normally acceptable temperature when used—or is used in a mix that contains enough ice or cool water—as part of the designed mix water—to bring the mix down to the desired temperature. (Liquid nitrogen also has been used to cool hot cement.)

When all you have is hot cement

Q. *We have found that we will have to use some hot cement to place a concrete dome on a water storage tank. This cement will be shipped directly from the mill to the project and there will be little time for it to cool after it comes out of the grinding mill. Is there apt to be a problem in using this hot cement?*

A. Hot cement will affect concrete only to the extent that the temperature of the cement influences the final temperature of the concrete, and that is very little. The problem can be controlled through use of normal hot weather concreting procedures. This is discussed in detail in a paper by William Lerch, "Hot Cement and Hot Weather Concrete Tests," published by the Portland Cement Association in 1955.

Has cement been stored too long?

Q. *We are storing and batching our own cement and aggregates and mixing our own concrete for a large job. Some portland cement has now been stored 5 months in the silos. There isn't any testing laboratory nearby and we wonder if there is a simple test to determine whether the cement is still good. It still flows freely and is not lumpy.*

A. You have probably already made the simplest test; the fact that the cement is free-flowing and free of lumps means that it is likely still good. Cement can be stored for an indefinite number of years if it is kept dry enough.

If you must make a more definitive test, your best bet would probably be to have a testing laboratory make a test of compressive strength on mortar cubes using ASTM method C 109. The strength should meet the requirements of ASTM C 150.

Algae in the mix water

Q. *There is no clear-running stream or other good source of mixing water near one of the sites planned for our portable batch plant, though there is a small lake which is more like a big pond. The trouble is that it contains a lot of green stuff that is reported to be algae. Can we use the water in making concrete?*

A. Not while it contains algae, because they cause great reduction in strength. It is reported that algae can reduce the rate of hydration, so for that reason you may not get the strength required. Algae can also cause a great amount of air to be entrained, and this could reduce the strength even more drastically. Another problem is that algae tend to deposit on aggregate particles and interfere with the bond between the aggregate and the cement paste, making the problem still worse.

Seawater for mixing perlite concrete

Q. *A steel tank is to be built in a remote area to hold asphalt. A stable insulating base is needed and it is proposed to make this base of perlite concrete of 190 psi compressive strength. Can ocean water be used in the perlite concrete mix? The only alternatives would be to truck the water in or drill a well.*

A. If the bottom of the steel tank can be protected adequately against corrosion from chlorides and other salts in the concrete mix, the ocean water should be satisfactory. The water may have a small adverse effect on strength and the time of set, but if the mix is designed with this in mind it should be possible to use ocean water.

Seawater as mixing water for concrete

Q. *Can seawater be used as mixing water in concrete?*

A. Salts in seawater may speed up the set of concrete and have also been reported to moderately reduce the ultimate strength. Another potential problem is efflorescence, or formation of salt deposits on the concrete surface. There are several accounts, however, of *unreinforced* concrete made with seawater performing well for 25 years or more. Because of the danger of rebar corrosion, it is not a good practice to use seawater in reinforced concrete and seawater should definitely not be used for making prestressed concrete.

Water—for sickness or health

Q. *When and why does water change its effect on concrete—from* harming *it, by too high a water-cement ratio, to* helping *it during curing? What goes on?*

A. In the early life of a concrete mix one role of the water is to provide workability. The higher the water-cement ratio (within limits) the more workable the mix. But also, in the fresh mix, the higher the water-cement ratio the farther apart the new hydrating cement gel filaments or crystals will spread themselves. (They distribute themselves fairly uniformly throughout the water-filled space.) If more water is added at any time before the concrete has set, it increases this water-filled space and so the cement gel (also called cement paste) will be less dense and the concrete of lower quality.

Once the concrete has set, the total volume of the cement paste has been established; and if water is *then* added it will not increase this cement paste volume in the way it did before. That is, it will not make the paste less dense and the concrete of lower quality.

Even though the concrete has set, the chemical reaction between the cement and water continues. The various cement compounds are still very slowly dissolving from the surfaces of the original cement grains. After dissolving, they react with the water and form more crystals within the same space, making the paste more dense. This means that anything that can be done to keep plenty of water available within the concrete and replace any water that has been used up in the chemical reaction will contribute to the densification of the cement paste, thereby improving the quality of the concrete.

In summary, too much water in the fresh mix will make the cement paste too low in density and weak. But after the concrete has set, the richness or leanness of the paste has been established; then not enough water for curing will rob the cement of its maximum opportunity to hydrate and make the concrete strong.

Reinforcing

Proper fabric designations

Q. *While studying a blueprint I discovered that in one place the architect specified 6 x 6—W1.4 x W1.4 mesh to be used in a 4-inch slab. Yet in another place in the same plan the architect called for 6 x 6—10 x 10 mesh. What is the difference between these two kinds of mesh? The service appears to be about the same.*

A. There is no difference between the two products specified. The architect used two different methods of designating the same welded wire mesh. He should have been consistent, using only the new 6 x 6—W1.4 x W1.4 designation. This is a designation established by the Wire Reinforcement Institute several years ago. It expresses the wire size in terms of hundredths of a square inch of cross sectional area for the convenience of the designer. "W1.4" represents a cross-sectional area of wire equal to 0.014 square inch. The new system of designation is discussed in the article "New Welded Wire Fabric Numbering System Explained," published in our October 1975 issue, page 454. For more detailed information, see the brochure, "How to Specify and Order Welded Wire Fabric," Publication TF501, Wire Reinforcement Institute, 8361 A Greensboro Drive, McLean, Virginia 22102.

Wire mesh at joints

Q. *Should wire mesh run through the bulkhead at construction joints to tie separate pours together?*

A. This decision depends upon whether the construction joint is intended to serve as a control joint. Control joints must allow horizontal movement of the slab so that uncontrolled random cracks don't occur at other locations. If the construction joint is to function in this manner the mesh should not run through the bulkhead. If it's to be a bonded construction joint, any reinforcement in the slab is continuous through the joint.

Uses for epoxy-coated rebars

Q. *Have there been any anticorrosion applications of epoxy-coated rebars other than for bridge decks and parking garage floors?*

A. The Fusion Bonded Coaters Association (FBCA) reports that the following uses have recently come to light:

- Reinforcing steel hoops for precast concrete silos used for salt storage
- Precast channel slab roof decks—considered particularly applicable to pulp and paper industry uses
- Hot water pipe encased in concrete floor to alleviate condensation in salt storage warehouse
- Reinforcement for earth retention—In addition to reinforcing earth, steel reinforcement strips are used in load-supporting structures, such as retaining walls, bridge abutments and dams
- Epoxy-coated welded wire fabric has also been used for retaining soil embankments and a similar product is used in the stems of precast double-T's for bridge decks and parking garages

- Epoxy-coated welded wire fabric has been used for horizontal joint reinforcement for masonry wall construction in a nuclear plant

The coating thickness on welded wire fabric varies from manufacturer to manufacturer because no ASTM standard has yet been adopted. However, a standard is currently in preparation and the FBCA and the Wire Reinforcement Institute have also joined forces to help develop specifications for using epoxy-coated welded wire fabric in general concrete applications.

Protecting cut ends of epoxy-coated bars

Q. *After epoxy-coated bars are sheared isn't it desirable or necessary to provide some kind of protection for the newly exposed ends? If so, can this be done on the job?*

A. It is necessary and it is standard practice to coat them in the rebar coating plant if they are sheared there, or on the job if the bars are cut on the job. An epoxy made for the express purpose of patching or touching up the coated bars is used.

Handle epoxy coated bars with care

Q. *One of our subcontractors has been installing epoxy coated rebars in the first deck of a parking garage. We've noticed quite a number of nicks in the coating. These vary from scratch marks to chips of 1/4 inch or more. Are there standard precautions that should be enforced to protect the coating? We'd like not to have this problem continue on the decks remaining to be built.*

A. It's mainly a matter of using care every step of the way by such methods as

- using nylon or padded bands for bundling the bars
- using nylon or padded wire rope slings
- lifting at the third points or using spreader bars
- storing bars on padded or wooden cribbing
- avoiding moving the bars by dragging
- walking on installed bars as little as possible
- taking care not to drop tools or heavy materials on the bars

Meantime, you should see to it that the damaged bars are touched up with an epoxy coating that is compatible with the original coating.

Tolerances on bar straightness

Q. *Is there a standard tolerance on straight (not bent) deformed rebars? A recent shipment of 50-foot lengths of Number 18 bars has a slight fixed bend that is definitely an inelastic deformation; the bars are not just sprung. Do we have any justification for rejecting them?*

A. ASTM reinforcing steel specifications A615, A616 and A617 make no references to such tolerances. Fabricating tolerances for bending deformed bars are given in the Concrete Reinforcing Steel Institute (CRSI) *Manual of Standard Practice,* 23rd edition, 1980, in Chapter 7, pages 7-4 and 7-5. These tolerances, however, refer only to the relatively short lengths of bars used for standard bends and there is no quantitative requirement for the tolerance of straight lengths of deformed bars.

It is common to require that "Bars shall not be kinked." Bars with a kink in them at the time of manufacture normally go to the scrap heap; those with slight bends or twists are ordinarily accepted. Nevertheless, when bars leave the rolling mill they are reasonably straight. It is possible for something to happen to them enroute that might bend them beyond the elastic limit and put a permanent bend or twist in them. However, we know of nothing in any standard specifications to establish a tolerance on the amount of such inelastic bending. Any requirements as to the straightness of bars would therefore have to be written into the specifications and contract documents by the design engineer.

Steel fibers that don't ball

Q. *Someone told us that fiber balling that occurs in mixing steel fiber concrete is a thing of the past. How has the leopard changed its spots?*

A. Balling seems not to be a problem with fibers that have any of several special shapes. Some of these are

- fibers with crescent-shaped cross-section and with crimpings running across them
- fibers that have their ends deformed and that are temporarily held together side by side by a water-soluble glue that allows them to break apart individually during mixing
- fibers that have widened ends

Such fibers also tend to anchor themselves well in the hardened concrete.

Polypropylene fiber floor as base for ceramic tile

Q. *I want to build a concrete floor to be covered with ceramic tile. I'd like to be sure that no shrinkage cracks appear in this floor later and crack the tile. Would polypropylene fibers do the job? What difference would the fibers make in the batching, mixing, and finishing operations?*

A. Polypropylene fibers have been used as crack arresters, and using them is one of several ways you might consider. (Others are using steel fibers, using welded wire fabric, or using a mix made of shrinkage-compensating cement.) The fibers are said to be easy to mix; usually they can be obtained in premeasured bags to be added at the rate of one bag per cubic yard of concrete. The manufacturer can recommend the length of fibers to choose for this application. The fibers don't affect the workability much but they tend to slow down the rate of bleeding, so this may delay the time when the slab can be troweled.

Chloride limit for concrete to be post-tensioned

Q. *Is there a limitation on chloride ions in concrete if the post-tensioning steel is located in ducts and will be encased in grease? Isn't that adequate protection for the steel?*

A. The Post-Tensioning Institute informs us that there is no difference in the limitation whether or not the steel is protected as you describe. Apparently the idea is to guard against the presence of chloride ions in case anything goes wrong with the intended protection.

Reinforcing steel specified at 72,500 psi

Q. *The plans for a foundation we're working on call for some 1¹/₄-inch-diameter, high-tensile, ribbed reinforcing bars with a minimum yield point in tension of 72,500 psi. Our suppliers know nothing about this kind of material, and the engineer could offer no help. We considered proposing a substitution, using some 150,000-psi threaded post-tensioning bars, but rising costs have forced us to take another look. How are we going to get the steel we need, and where does such a requirement come from anyway?*

A. We turned to David Gustafson, technical director of the Concrete Reinforcing Steel Institute (CRSI), for help with this inquiry. He speculated that someone may have been working from a West European design, since *ribbed* is the preferred European term for what U.S. standards call *deformed.* This idea gets further support when we convert 72,500 psi to metric units. Rounded off, it comes out to be 500 MPa (mega-pascals), a fairly common grade in Europe.

The current ASTM specification for billet steel reinforcing bars A 615-87 does include Grade 75 (75,000-psi minimum yield strength) bars, but only in #11, #14, and #18 sizes. You could substitute a #11 Grade 75 bar for the one your plans call for; it would be a little larger and stronger. However, because of limited demand for this grade in the United States, it is not widely available as a standard product and you may have to do some hunting. Rebar sizes #14 and #18 are not commonly stocked by all producers or fabricators. The prudent designer intending to use Grade 75 in any size should consult suppliers directly or through local constructors.

Can mesh be too high?

Q. *I frequently hear warnings about leaving wire mesh too low in a concrete slab on ground. I'm told the mesh won't keep cracks closed tightly if it isn't high enough. Is there any problem if it's too high, say 1/2 inch to 1 inch from the surface?*

A. If wire mesh is too close to the surface it may corrode, especially in exterior applications where the concrete won't be continuously dry. Generally, the minimum cover for wire mesh should be 1 1/2 to 2 inches.

Mix design

Water-cement ratio

Q. *We repeatedly hear that the strength of concrete depends almost entirely on the water-cement ratio. Yet I cannot understand why two mixes, one containing four bags of portland cement per cubic yard and the other containing six bags per cubic yard, could have the same strength. The second has 50 percent more cement than the first. Isn't the second mix really stronger at the same water-cement ratio?*

A. The two mixes will be very close to the same strength. Suppose that you make two mixes of cement and water only. One has four bags of cement and the other six. Both contain, let us say, half as much water as cement by weight. When they have set and cured they will both have the same strength, though one will occupy a larger volume than the other.

Now mix two more batches exactly like the first ones. Into each one add enough saturated surface-dry aggregate to make a total volume of one cubic yard. Obviously one mix will take a little more aggregate than the other. Now, when they have cured, the weakest link in both masses will be the cement-water portion: the cement paste. The cement paste will determine the strength. But we've noted before that both pastes have the same strength; consequently both concretes will have the same strength.

There are, of course, some slight differences in strength of the two mixes, because there is less aggregate surface to be bound in one mix than the other, but the difference is minor.

There is a qualification to the above discussion: The explanation does not apply to concretes in which the aggregate is weaker than the cement paste. This is likely to be true, for example, of concretes made with perlite, vermiculite, or polystyrene aggregate.

Color uniformity in vertical slipforming

Q. *We are slipforming some 80-foot-high lime silos for a sewage treatment plant and the sewage treatment authority wants the appearance of the concrete to blend with the nearby natural rock of the environment. We want to do this by painting but the owner insists on integral color in the concrete.*

We don't think we would be able to keep the concrete color uniform. The slipforming is to be done during summer at about 120° F daytime and 100° F nighttime temperatures. The mix is designed for 4-inch slump but specifications permit slump up to 6 inches. We expect that as the job progresses we will probably have to change the amount of retarder, ice, and water between day and night conditions and we think this will add to our problems with color and make it impossible to keep it uniform. Our concrete supplier agrees. Do you have any advice?

A. The variables inherent in the slipforming operation—changes in water-cement ratio, temperature and quantity of retarder—combined with inevitable changes in mix proportions and rate of slide, all foretell nonuniformity of color. Furthermore, there is no way to compensate, from hour to hour or shift to shift, for these differences, because the final color won't be known until the concrete has been cured and dried—a matter of weeks, at least. If you can't get the owner to give up the use of integral color, you might ask the color manufacturer if he is willing to make a statement for the owner about the limitations of color in such an application.

Measuring yield

Q. *What is meant by yield of concrete?*

A. Yield is the volume of concrete produced from a mixture of known amounts of cement, water, aggregates, and admixtures. Concrete is batched by weight but sold by volume so a unit weight test is usually conducted when it is necessary to calculate yield. By finding out how much a cubic foot of concrete weighs and dividing that number into the batch weight, the batch volume or yield is determined.

Slump of concrete for residential walls

Q. *The contractor who plans to build the foundation walls for our house makes a practice of using concrete with a slump of 6 ± 1 inch. I've argued that this would be too wet, but he says that his trade association approves this practice and that stiffer concrete can't be consolidated properly. He says he needs this kind of slump to avoid honeycombing. He also says he orders a mix that is designed to give the required strength at this slump. Does he know what he's talking about?*

A. If the contractor does order concrete to achieve the required strength *at this slump*, he is taking the right approach. In recent years the Poured Concrete Wall Contractors Association of America has advocated wetter mixes for residential wall construction in order to facilitate consolidation by puddling and without a vibrator. This approach is also being taken by ACI Committee 332 in its "Guide to Residential Cast-in-Place Concrete Construction," with the provision that the concrete be proportioned to meet the required strength at this slump and not some lower slump. The contractor should use job practices that prevent the segregation that can occur more easily in wet mixes.

Mixes for exposed aggregate walks

Q. *Is there a standard mix to use for exposed aggregate flatwork if you do not want to seed the top of the concrete with a special aggregate?*

A. No, there is not a standard mix; but you should design the mix or choose the mix with the desired end result in mind. This means that you will probably want to consider what maximum size of aggregate is best for the effect you want to achieve. Pea gravel (3/8-inch maximum) is a popular size for the purpose. It is often helpful to use a single size of coarse aggregate (1/4 to 3/8 inch, or 3/8 to 3/4 inch, for example) rather than a graded aggregate. This tends to give a good appearance. If you use anything other than a standard mix you should make sure it has the proper workability; if not, you may have to increase or decrease the sand content to make it suitable for placing and finishing. All of this presupposes that the mix will be designed for adequate strength (preferably 4000 psi if for outdoor use in a freezing climate), will have a slump of 2 to 4 inches and, if to be exposed to weather, will be air entrained.

Designing a mix for flexural strength

Q. *Our ready mixed concrete company has received a request for concrete with flexural strength of 670 psi. We have never had this kind of order before. Can you tell us what equivalent compressive strength such a concrete represents?*

A. This sounds like an order for paving concrete, which is commonly specified in terms of flexural strength. Flexural strength can be related to compressive strength of concrete by the following formula:

$$f_r = 9 \sqrt{f'_c}$$

This says that the flexural strength is roughly equal to 9 times the square root of the compressive strength. On this basis a flexural strength of 670 psi would be equivalent to 5540 psi compressive strength. The formula is not exact because it is influenced by the type of aggregate. On paving jobs the modulus of rupture is determined by ASTM C 78 "Test Method for Flexural Strength of Concrete (Using Simple Beam with Third-Point Loading)," and cylinders for compressive strength are not likely to be made.

Concrete piles in acid soil

Q. *A structure has been designed to be built on 12-inch-diameter piles on a 20-foot grid on a low-lying island about 20 miles upstream from the mouth of a river. Tests have shown the soil to be quite acidic—within the range of about 3.8 to 4.6. We're concerned about acid attack on the piles. Is there a concrete mix that would be durable in such an environment? If not, what kind of protection would you advise for members that can't be reached for recoating?*

A. There's evidence that ordinary concrete of good quality would be durable with no coating or membrane. In a letter to *Concrete* magazine (British), February 1986, page 23, T. P. Lees of the Cement and Concrete Association reports findings that concrete that has been in soil at pH 3.5 to 4.5 for 20 years showed "no sign of any sort of attack." He also quotes another investigator, Marjory Eglinton, who has said "there is no evidence of structural weakening of mass concrete exposed to . . . flowing water containing [carbon dioxide] and humic acid and no evidence of damage to concrete placed in acidic ground conditions when free water is absent." T. P. Lees believes that where the pH isn't caused by industrial wastes but by native acids, where the pH is 3.5 or more, and where there is no *flow* of water across the concrete, there isn't any cause for concern. There may be cause if the pH is less than 3.5. Natural acidity rarely falls lower than this, so a lower pH usually indicates an outside source of acid.

A subsequent letter to the same magazine (April 1986, page 52) from Guy Brennan, of The Concrete Pipe Association in England, supports the views of Lees and reports good performance by precast concrete sewer pipe that had been subjected to untreated effluent from a plating plant where the pH alternated between 1 and 12 in three-hour cycles for some unspecified length of time. When examined, the concrete surface had been well etched but the body of the concrete was sound.

In another instance concrete specimens had been obtained from test borings that penetrated some old foundation beams in the soil on the site of an old gas plant. From the context of his discussion the soil must have been acidic. These stained and dirty. concrete fragments were opportune test pieces because they already had a history of exposure to soil acids. Examination showed the bodies of the test pieces to be unaffected and their surfaces to be sound and firm. Brennan recommended that there should be no problems in specifying the use of concrete structures in this area. Unfortunatley, his published letter doesn't report the pH of the soil. He acknowledges that extreme cases undoubtedly exist where "some of the more virulent moorland waters (with) just the right type of acid at exactly the wrong concentration might affect some concrete." However, he is convinced that "well constructed (concrete) products that satisfy today's demanding specification should have no difficulty in surviving such an environment."

According to The Concrete Pipe Association, those precast concrete pipes and other drainage products that meet the specification contained in British standard BS 5911 are resistant to most moorland waters. Many of these waters contain humic or carbonic acids. Concrete of the quality required can be characterized as having high cement contents and low water-cement ratios, being highly compacted, and having low porosity and low absorption.

Nailable concrete

Q. *We've had a recent inquiry about how to make nailable concrete. We recall that sawdust concrete was promoted for this purpose at one time. Is there anything new on the subject?*

A. Very lightweight concretes accept ordinary nails readily. Normal-weight concretes can be nailed with powder actuated tools.

At least 40 years ago sawdust concrete was promoted for its nailability (see "Waste Materials in Concrete," September 1971 issue, page 372.) Perlite concrete, vermiculite concrete, cellular concrete and concrete made with foamed polystyrene beads, which in this country are usually made in the density class of about 20 to 50 pounds per cubic foot, will all accept nails. The same concretes made at higher densities will undoubtedly do so also. We know of no standard method, however, of testing nailability of concrete, nor any specifications for nailability.

Normal-weight concrete can routinely be nailed with powder actuated tools (see "The ABC's of Powder Actuated Tool Fastening," March 1976 issue, page 113). The method can apparently be used successfully with concrete in the compressive strength range of about 1400 to 5000 psi.

A cause of unsatisfactory shotcrete

Q. *The walls of some tunnels between several of our plant buildings were made by wet-mix shotcreting. The shotcrete material was brought to the job from a ready mixed concrete plant and fed into the hopper of the shotcreting equipment. In order to minimize delays the trucks were required to arrive well ahead of time, and much of the time there were two trucks at the job site. The walls are not satisfactory. In some places they have a hollow sound when tapped, and the shotcrete appears to be of generally low quality. Since shotcrete is normally expected to be of high quality we are puzzled.*

A. It seems possible that your problem is the result of some of the shotcrete being too old when it was used. Forty-five minutes after mixing has generally been considered to be the maximum elapsed time for placing wet-mix shotcrete, whether it is mixed on the job or supplied by a ready mixed concrete producer.

It is interesting to note that in August 1979 the Gunite Contractors Association sent out a news bulletin alerting the industry to this time limitation. They report that in a number of instances in the preceding months they had investigated reports of wet-mixed shotcrete of extremely low quality. In every case they found that the delivery tickets showed that "time spans varying from 1 hour 10 minutes up to 4 hours 30 minutes had elapsed after the sand and cement had been introduced into the transit mix truck and until the material was placed."

Even 45 minutes may be too long, the Association points out, if "there is excessive moisture in the aggregate, a very high ratio of cement to sand, high ambient temperatures and low humidity." On the other hand the reverse of all these conditions could make 1 hour 15 minutes acceptable. The aggregate grading, fineness modulus and capability of the mixing equipment might also affect the allowable time.

Unfortunately there is no good way to detect in the field whether the cement in the mix has begun to set, but the Association suggests that it is better to discard mixes that are too old than to suffer the penalties of an unsatisfactory job.

The quick fix

Q. *Our company sold a 10x20-foot sign for use in southern California. This is mounted on legs made of 12-inch-square steel tubes. Another company contracted to set the sign in place. With the legs in 2-foot-diameter holes 7 feet deep, they mixed some concrete and dumped some flake calcium chloride into the mix directly from the bag. Then they placed the concrete in the holes around the steel legs. Are the legs going to corrode quickly, break up the concrete and make the sign fall down?*

A. It probably seems attractive to cause concrete in sign postholes to set rapidly, and it is possible that your subcontractor may even have added an overdose of calcium chloride. That would permit getting off the job quickly after placing the concrete.

Whether there was an overdose or not, it is always considered bad practice to add solid calcium chloride to the mix. ACI 212.2R, "Guide for Use of Admixtures in Concrete," says that "calcium chloride should be added to the concrete in solution form" and it explains how to make up the solution.

The presence of chloride ion can contribute to corrosion of steel in concrete if both oxygen and moisture are available. Consequently, corrosion could begin where the steel enters the top of the concrete or at some place where the concrete becomes cracked. One cannot say for certain that you will have quick, disastrous corrosion—or even slow disastrous corrosion. If your environment is very dry it could greatly prolong the life of the legs. You might just keep an eye on the bases of the legs to see what is happening to them; or you might demand that the work be redone on the basis that the concrete was not mixed according to acceptable procedures.

Mix proportions for the concrete you buy

Q. *If I specify strength, air content, and slump when I order concrete, is the ready-mixed concrete supplier obligated in any way to tell me the mix proportions? I haven't been getting this information, and though I don't absolutely have to know what material and how much of each material is in the mix, I'd feel a lot better if I had the information.*

A. The information might not only make you feel better, but might eventually lead you toward participation in more careful selection of mixes. ASTM C 94, Section 5.3.1 provides that "when the purchaser requires the manufacturer to assume full responsibility for the selection of the proportions for the concrete mixture . . . the purchaser shall also specify " 28-day or other compressive strength. Section 5.3.2 says *"At the request of the purchaser* [italics ours] the manufacturer shall, prior to the actual delivery of the concrete, furnish a statement to the purchaser, giving the dry weights of cement and saturated surface-dry-weights of fine and coarse aggregate and quantities, and name of admixtures (if any) and of water per cubic yard or cubic meter of concrete that will be used in the manufacture of each class of concrete ordered by the purchaser. [The manufacturer] shall also furnish evidence satisfactory to the purchaser that the materials to be used and proportions selected will produce concrete of the quality specified." All you should have to do is ask.

Formwork

2

Concrete

Formwork pressure calculations

Q. *American Concrete Institute formulas for lateral pressure of concrete (ACI 347-78, Section 2.2) show that pressure is inversely related to the temperature of the concrete in the forms. However, the temperature of the concrete is rarely (if ever) monitored in the forms. In cold weather, concrete may be heated to 60 to 70° F when mixed. It is then placed in steel or wood forms which are at air temperature. The placement from bottom of form to top can easily occur over a period of 1 to 5 hours. There is an initial heat loss to the cold forms and a continuing heat loss through steel or plywood. Can one assume an appropriate concrete temperature for the formulas or should ambient temperature be used for pressure computation?*

A. There is nothing in the ACI 347 Standard itself that answers this question. Some heat is generated in the concrete as the cement hydrates. However, the effect of this heat varies greatly, depending on job conditions. Form material, thickness of concrete cross section, wind, condition of forms and type of external protection all have an influence. The effect of cold forms and heat loss through the forms would slow down the rate of hardening, thus keeping the internal pressure on the forms high for a longer period of time.

This is the same effect a retarding admixture would have, though not necessarily of the same magnitude. For this reason it might be well to follow the recommendation: "Where form temperature is less than that of the concrete, an effective value of temperature less than that of the concrete in the forms should be used in the formulas."

This is an adaptation of a recommendation given in Section 2.2.2 (f) of the Standard. (That section concerns retarding admixtures but not temperatures). In effect this advice counsels making an informed guess about how much lower a temperature value should be used in the formulas. One could assume an appropriately low concrete temperature. It would be better still to measure it. Or, to be completely conservative, the ambient temperature could be used, provided it is 40° F or more. If the ambient temperature is below 40° F and the concrete is not protected against cold weather, you could calculate the hydrostatic pressure in pounds per square foot by multiplying the height of the concrete in the forms by 150.

There may be further problems if use of fly ash or retarding admixtures is intended. The use of these materials in cold weather should be seriously questioned because they prolong the rate of setting. If they must be used, extend the length of time before the forms are stripped.

The ACI pressure formulas were developed on the basis of actual experience, using information for temperatures which did not go below 40° F. Winter concreting recommendations such as those published by ACI Committee 306 require concrete to be 40° F or higher when placed and to be protected from freezing for some minimum period of time. Without this required protection, concrete conceivably could freeze at the time of placement (relieving lateral pressure of course), then thaw later and overload the forms or lead to collapse of the structure. This actually happened during the early days of concrete building construction, as Jacob Feld explained in the article "Temperature Effects on Completed Work and During Construction," *Concrete Construction*, March 1964, page 61.

Safety factor for form ties

Q. *Although we aren't having any trouble with our form ties, I wonder how safe they really are? Is there a standard factor of safety, and who decides what it is?*

A. Design professionals plan structures using established safety factors, many of them clearly defined in codes and design standards. In construction, there are many cases where safety factors have not been defined so precisely by outside authorities. Much has been left to the judgment and experience of contractors and materials suppliers. Load ratings for formwork accessories sometimes fall into this category. Frequently the manufacturer determines the ultimate or breaking load for a particular product, assumes an appropriate factor of safety, and arrives at a load rating for the product. Typically, manufacturers have assigned form ties a safety factor of 1.5. Thus, if a standard snap tie breaks at a load of 4,500 pounds, it is given a safe load rating of 3,000 pounds.

These customary load ratings must be reexamined in the light of recommendations adopted by the American Concrete Institute (ACI) Committee 347, *Formwork for Concrete.* ACI 347's report shows minimum safety factors of 2 for almost all formwork accessories—ties, anchors, and hangers. Some form anchors are even required to have a safety factor of 3.

To comply with these recommendations, you'll have to read the fine print in the catalogs. Usually the safety factor used in arriving at the load rating is given. Often it is 1.5, although some sellers provide ratings based on both 1.5 and 2. Regrettably, we have even seen one catalog that showed safety factors as low as 1.25.

To convert a load rating from a safety factor of 1.5 to 2, multiply by 0.75. Or if the catalog gives an ultimate load, just divide that by 2.

For embedded anchors, you also must check the concrete strength needed to develop the capacity of the anchor.

There is another safety concern for the form designer. Form ties are by definition tension members. They hold forms together by resisting lateral pressure of concrete, and published load ratings are generally for tension loads only. Unfortunately, real-world applications don't always conform to the definition. Bending and shear loads applied to form ties which weren't designed for them have been involved in some tragic failures. Manufacturers need to provide information on shear capacity of their products. Users need to be alert to field conditions, designing for loads other than tension where necessary.

We mention the ACI recommendations here because for some years American Concrete Institute, through its formwork committee, has been recognized as a consensus-standards maker for formwork. The American National Standards Institute (ANSI) regulations for safety in concrete and masonry work cite ACI. Existing Occupational Safety and Health Administration (OSHA) regulations for concrete refer to ANSI standards, thus indirectly citing ACI. The upcoming OSHA revision appears to be silent on specific factors of safety, but cites ACI 347 as a helpful reference.

Effect of superplasticizers on form pressures

Q. *To minimize bugholes in exterior walls we are considering using a high-range water reducer to give us flowing concrete. Is there a way of calculating how much higher the formwork pressures would be?*

A. According to "Superplasticizing Admixtures in Concrete," a 1976 publication of the Cement and Concrete Association (England), wall and column forms to receive flowing concrete should be designed for full liquid pressure plus an impact surcharge of about 210 pounds per square foot if concrete is discharged from a height of 6.5 feet or more. The publication adds that if there are any cleanout openings in the forms, special care should be taken to see that their covers are well anchored.

Wood for formwork

Q. *Is it better to use air-dried or kiln-dried wood for formwork?*

A. The general statement given in the American Concrete Institute's *Formwork for Concrete* is that partially seasoned stock is preferred since fully dried lumber swells excessively when it becomes wet, and green timber will dry out and warp during hot weather, leading to misalignment or uneven surfaces.

Looking deeper into the matter we consulted George Oelrich, who has spent much of his life working with and studying wood. He commented that air-dried wood is by far more stable, citing its use by makers of fine musical instruments and furniture, who consider stability critical.

Oelrich went on to say, "If this is true, then why use kiln dry wood? Kiln drying, though costly, produces a usable board in a shorter time (days) than air drying (weeks). Since time is money, economically it is better to kiln dry than to air dry. In most instances the effects on the wood (increase in density, induced stress from cell shrinkage, brittleness, etc.) do not affect the end use of the wood as long as moisture is not reintroduced. . . .

"Unfortunately not only do we reintroduce moisture in concrete forming, we do so under pressure. If there are any unrelieved stresses in the wood, it may twist, turn, cup, bow, or warp in or on the form. All this takes place while the concrete is still plastic. The resulting concrete will mirror what has taken place."

This can be a matter of considerable concern in forming rustication grooves and other sensitive details in architectural concrete. So Oelrich suggests using air-dried wood chamfers and rustication strips, verifying with the local supplier which kind is available. Cost differences between the two should be small.

Boat patches

Q. *How do you avoid the little shapes that look like pointed ovals on concrete surfaces? Would a paint or coating on the plywood forms take care of them?*

A. You must be referring to plywood form marks from what are usually called boat patches in the plywood. The plywood manufacturer stamps out these boat shapes around knots in the surface ply and then inserts precut patches of the same size and shape. The boat patches make the surface of the plywood smoother than if it contained the knot, but the patches aren't pretty. B-B Grade plywood, or any other grade that contains boat patches, should not be used where this kind of imprint on the concrete surface will be objectionable.

Do ribbed form liners limit aggregate size?

Q. *Your article on decorative form liners (May 1987, page 431) failed to comment on suitable concrete mixes. If we have ribbed form liners, can we expect problems with aggregate being too large to fit into the ribs?*

A. Apparently this must not be a common problem. We consulted manufacturers' literature, books on formwork, and several American Concrete Institute (ACI) guides and standards, but found only one source that contained a specific warning. In a recently-issued jobsite guide, Greenstreak says, "For ribbed textures the aggregate should be smaller than the width of the rib. Oversize aggregate can cause honeycombing and chipping on the ribs."

Even more critical to good results with ribbed form liners is proper consolidation of the concrete by vibration. Several sources stress the need for added vibration because the textured liner has so much more surface area than a flat form. ACI recommends depositing concrete in layers 12 to 18 inches thick and vibrating one lift at a time, extending the vibrator at least 6 inches into the previous lift at each insertion point. The vibrator is held stationary 10 to 30 seconds or until consolidation is judged to be complete. Then it is withdrawn at about 3 inches per second. Symons Corporation application guides suggest that, with a deeply textured form liner, the penetration into the previous lift be 6 to 12 inches and the rate of withdrawal be reduced to 1 to 2 inches per second.

Avoid butt joints between form liners

Q. *We like the look you can get with textured plastic form liners, but we always seem to have trouble fitting them together or matching up ribs when we place a second lift. How can we do a better job?*

A. You're not alone with this problem. In general, butt joints between either horizontal or vertical liners should be avoided because it's difficult and time-consuming to match liners so they appear continuous. Some ribbed liners are made with interlocking flanges, but this helps with joints in one direction only. Some of the elastomeric (rubbery) liners that have to be tightly and accurately mounted on form panels are flexible enough to permit satisfactory butting. This helps, but only within the area covered by a single panel.

The best solution, which frequently involves the cooperation or consent of the architect, is to use a rustication strip or groove former to accentuate rather than hide the joint. The rustication strip should also be used at all cold joints in architectural concrete (see photo). Certain liners having random patterns cannot be matched either horizontally or vertically. A rustication strip is always advised at joints in such patterns.

If butt joints are unavoidable, be sure to prevent grout leakage with compressible foam tape, caulking, or other material suitable for the particular liner. For liners with

Liner texture variations and the high compressive strength of most liner materials make it hard to accurately lap a form liner over a textured surface in a previous pour. Thus, nontextured areas or recessed divisions (rustications) between pours are preferred as shown in this finished wall.

a strong vertical pattern, make the joints along the main features of the liner. Position the joints so they fall in recessed or shadow areas of the concrete where they are less conspicuous.

Soy bean oil for form oil

Q. *Our local ready mixed concrete association is interested in the use of soy bean oil as a form release agent for architectural concrete. Is soy bean form oil ever specified for this use and does anyone market soy bean form oil commercially?*

A. We have not been able to find anyone who uses soy bean oil for this purpose, though so many oils have been used as form release agents that it could well have been used on some jobs. Castor oil has been used as a form release agent for rubber form liners that might swell from use of petroleum oils.

Expanded metal bulkheads

Q. *Our company is building an 8-foot-thick mat foundation for a sewage treatment plant. We want to use expanded metal as a forming material at the vertical construction joints. Since the mat is heavily reinforced, there are many penetrating reinforcing bars. We think it would save us time and money to use expanded metal instead of building solid bulkhead forms, but the engineer refuses to approve this unless we can show that it is "approved by ACI." Can you help us?*

A. We were unable to find any reference—pro or con—in published American Concrete Institute (ACI) standards and committee reports. However, we did learn that a photograph of expanded metal was included in *Formwork for Concrete* (ACI Special Publication No. 4), and an accompanying description was added about the use of expanded metal as a bulkhead form.

In current books, we found that John Richardson of the English Cement and Concrete Association mentions it in *Formwork Construction and Practice* in this way. "Another method of forming joints is by using expanded metal on supporting frames. The expanded metal forms a textured surface and when torn away from the surface at the joint provides an excellent key for further work." It is likely that Richardson refers to a generic form of expanded metal, rather than the ribbed products manufactured in the United States specifically for forming operations. It's hard to imagine ripping off the metal at a joint that looks like the one in the photograph, nor would it be necessary.

Ribbed expanded metal bulkhead has been wire-tied to vertical bars in a nuclear powerplant in Georgia. Horizontal bars have been threaded through it.

PHOTO CREDITS: *Expanded Metal Corporation*

Mortar from the concrete filters through holes in the expanded metal mesh. Since it hardens to form a rough, toothy surface to which the next concrete placement can bond, no scabbling is necessary.

There is a growing precedent for acceptance of expanded metal as a bulkhead form, and manufacturers can supply information on enough existing installations to satisfy most of the skeptical specifiers. A full description of the material and ways to use it were presented in *Concrete Construction*, July 1983, page 552.

Do expanded metal bulkheads cause corrosion problems?

Q. *Your Problem Clinic item on expanded metal bulkheads (November 1986) is interesting but I'd still like to have more information. Are there jobs that have been in place for 15 or more years so the long-term performance can be evaluated? Can zinc in the galvanized expanded metal product form a corrosion cell with the rebar?*

A. We talked with two manufacturers of expanded metal products. Neither had long-term performance data for their product when it's used in bulkheads. We then contacted several concrete industry sources. John Richardson, a concrete specialist from the United Kingdom, reports that expanded metal is used there for both stay-in-place forms and bulkheads. He says it was used by the acre on the Thames barrier and that there were no reports of corrosion problems. He cautions that there must be adequate cover over any metal left embedded in concrete.

Both design and construction people in the United States have told us of successful use of expanded metal for form bulkheads in nuclear power plants and water retaining structures. Some of this work has been in place for more than 15 years with no problems reported.

Expanded metal has been used in ferrocement boats for many years. Martin Iorns, a ferrocement consultant from Sacramento, California, told us that corrosion isn't a problem in boats. Even when hull abrasion scrapes off the mortar cover and exposes expanded metal to salt water, rusting doesn't develop enough pressure to cause further spalling of surface mortar. This is because there isn't enough steel mass to generate a significant volume of corrosion products.

Easier to strip foam if aggregate can be exposed

Q. *We have been using extruded polystyrene foam form liners on tilt-up panels. During stripping this material breaks up and some of it usually sticks to the concrete. This has to be peeled off piece by piece, a laborious job. Is there a way to eliminate this sticking problem and the time-consuming cleanup?*

A. There is if you would be interested in exposing some aggregate at the same time. Gayle Johnson of Ray Hoopes Concrete and Stan Hanson, developer of the H & H Industrial Park in Phoenix, Arizona, report that they found an easy solution to their sticking problem. Instead of the usual form release agent they applied a specially formulated water-based retarder to the surface of the polystyrene form liner. This delayed the hardening of the surface layer of concrete and permitted easy stripping. After stripping, the retarded mortar was washed from the panel surface with a hose, exposing aggregate (which had been specially selected) and producing pleasing shadings and textures. This produced economically an unusual combination sculptured-and exposed-aggregate finish.

Slab form removal in multistory work

Q. *We frequently are asked to strip slab formwork based on compressive strength, as determined by test cylinders, reaching 80% to 85% of the 28-day specified compressive strength. Often we get this strength 3 days after placing the concrete.*

However, we notice that the American Concrete Institute (ACI) "Guide to Formwork for Concrete" says (Section 3.7.2.1) that when forms are stripped there must be no excessive deflection or distortion. How can we be sure there will not be excessive deflection? And what's excessive anyway?*

A. As you point out, we have two concerns in deciding when forms can be removed—strength and longer term serviceability. The structure must safely carry its own weight and any applied loads; it must also maintain its shape without unsightly or disabling deflections.

The answer to your question depends a lot on the use of the structure and the kind of interior finishes and partitions it will have. The structural engineer usually is in the best position to decide how much deflection is allowable, based on code requirements, as well as his or her experience. Because concrete deflection is more severe when loads are applied at very early ages, a time limit before forms can be removed is frequently specified. The exact time depends on the structure's design and on the construction methods.

As to strength and safety, in some flat plate buildings where several levels of shores and reshores may be used, it's possible for the load on a slab to build up to two or more times the service load for which the slab was designed. If the contract documents specify the conditions for form removal, the Occupational Safety and Health Administration (OSHA) permits the contractor to follow these specifications, usually based on testing of the concrete (see *Concrete Construction,* December 1988, page 1079). If contract documents don't cover this, you must test the compressive strength of the concrete and determine that test results "indicate that the concrete has gained sufficient strength for the concrete to support its weight and superimposed loads." If contractors do not have an engineer on staff, they may want to consult one for help with this determination.

Notice that the new OSHA regulations do not specifically require *cylinder* tests. Any "appropriate ASTM standard test method" may be used. This opens the way for in-place testing but doesn't rule out cylinders if you want to use them.

Meeting OSHA requirements enables you to remove forms safely. Preventing undesirable deflection in the slabs is more difficult, particularly if you want to maintain a fast construction cycle. *Backshoring* and *preshoring* methods explained in the ACI 347 Guide and in an earlier *Concrete Construction* article (see March 1986, page 307) help control deflection when construction moves at a fast pace.

* Published by the American Concrete Institute, P.O. Box 19150, Detroit, Michigan 48219

Forms and shores cannot be removed until concrete is strong enough to support its own weight and any applied loads such as the construction materials seen here. Reshores, backshores, or preshores may be used to help distribute loads among several interconnected slabs, thereby speeding up the construction cycle.

Tilt-up uses *trompe l'oeil*

Q. *What does a French term like* trompe l'oeil *have to do with concrete? I've seen it several times recently. And also how do you pronounce it?*

A. The best advice we've had on pronunciation is:

trom-ploy

with accent on the first part of the word. As you already know, it is a French expression. Literally it means "fool the eye." In painting, it refers to the representation of an object so exactly as to deceive the eye. The technique was used by the ancient Greeks; Zeuxis, for example, reportedly painted such realistic grapes that birds tried to eat them.

This tilt-up building in California is made of simple, flat panels painted to resemble deeply coffered units—an example of trompe l'oeil *in current construction overlay.*

PHOTO CREDITS: Dayton Superior Corporation

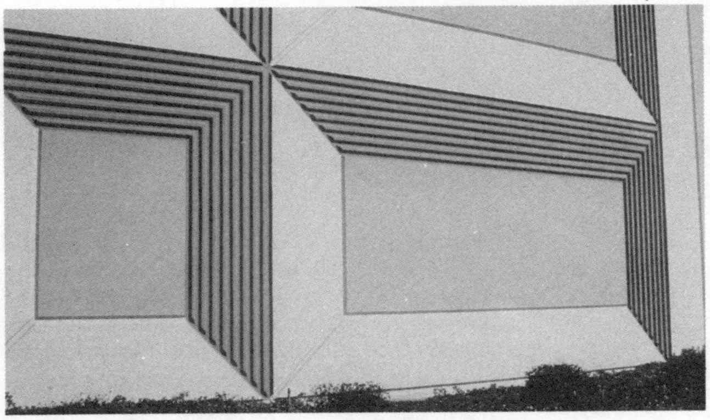

More recently flat murals and ceilings have been painted with architectural detail to suggest complicated three-dimensional elements. Tilt-up designers and builders have found this an economical way to add a dramatic sense of depth to panels cast flat at the building site. Sometimes wood rustication strips are used to cast grooves, perhaps an inch deep, that become part of the painted *trompe l'oeil* as shown in the accompanying photographs provided by Peter Courtois. See also the tilt-up article in *Concrete Construction,* September 1987, page 843.

A safer practice

Q. *In high-rise construction we never remove our shores until we're satisfied by tests that the floor has achieved sufficient strength. In recent years we've switched from cylinder tests to cast-in-place cylinders that are removed from the floor at the time of testing. Nevertheless, being aware of structural failures, I'd like to know whether there is any other pre-*

caution we could adopt that would make me feel just a little more secure against a catastrophic failure.

A. One contractor with 40 years experience tells us that when he strips the lowest set of shores he always leaves about 25% of them in place—just as the kind of precaution you're inquiring about. This means that he has the equivalent of $2^1/_4$ floors of shoring instead of 2, but he always feels safer. When he strips the next floor he takes out 75% of the shores and goes down to the lowest shored floor and removes the last 25%. It's a little bit of a nuisance to work on two floors instead of one, but it makes him and his crew feel better, and it may pay off in preventing a failure. In case it is expected that loads will be applied to any part of the top floor he also carries some point shoring down three levels below that load.

No error in these published times for removing support

Q. *I think there is a typographical error in "Recommended Practice for Concrete Formwork, ACI 347-78 (Reaffirmed 1984)." Section 3.6.2.3 contains a table showing minimum periods of time that formwork and supports should be kept in place under ordinary conditions. The time periods are longer in the column titled "where design live load is less than dead load" than in the column "where design live load is greater than dead load." Haven't these column headings been accidentally reversed? Otherwise how do you explain that a comparatively weaker structure is expected to carry a greater live load than a stronger structure?*

A. This is a thoughtful question, but the column headings actually are correct. Essentially this section of ACI 347-78 says that, where the *design live load* is greater than the dead load, you can remove forms sooner than if the design live load is less than the dead load. This is based on the assumption that the design live load won't be there during construction, but that some percentage of the strength required to support it *will* be. Stripping can be started safely when the slab or beam is able to support itself. The slab or beam can use for this purpose some of the capacity that the designer planned for live load. Let's compare two cases:

Slab A: live load **50** pounds per square foot
dead load **100** pounds per square foot

Slab B: live load **150** pounds per square foot
dead load **125** pounds per square foot

Assume, conservatively, that the concrete will attain 60% of design strength by age 4 days. Then Slab A can carry 0.60 x 150 or 90 pounds per square foot. However, the slab itself weighs (has a dead load of) 100 pounds per square foot. So it's overloaded unless we change the safety factors.

Slab B can carry 0.60 x 275 or 165 pounds per square foot, and it weighs 125 pounds per square foot. It's safe, using the same safety factor as for the permanent structure.

Admittedly these numbers are fictitious, but maybe they can give you the idea of what is behind the wording of the standard you asked about.

Differences between reshoring and backshoring

Q. *We have heard the expressions* reshoring *and* backshoring *used quite often. What's the difference between the two methods?*

A. Peter Courtois, former chairman of American Concrete Institute Committee 347, Formwork for Concrete, gave us the following response.

We'll preface the discussion by first using the definitions from an unpublished report of the American Concrete Institute's formwork committee.

Shoring: System of vertical or inclined supports for forms; may be wood or metal posts, scaffold-type frames, or various patented members.

Reshoring: Shores placed snugly under a stripped concrete slab or structural member after original formwork and shores have been removed from a large area, thus requiring the new slab or member to deflect and support its own weight and existing construction loads prior to the installation of reshores.

Backshoring: Shores placed snugly under a stripped concrete slab or structural member after the original formwork and shores have been removed from a small area without allowing the slab to deflect or support its own weight or existing construction loads from above.

The original shores must be installed plumb and wedged securely so that each carries its share of load. When the original shores and formwork are removed, the slab or beam is required to support its own weight and any other loads that may have been applied. Reshores are provided to distribute any further applied loads among several slabs that are stronger (older), thereby helping to maximize the reuse of the formwork.

When placing reshores, one must be sure not to remove the normal deflection of the floor above. The reshore is simply a strut and should be wedged only firmly enough to hold it in place. Do not set wedges with a sledgehammer, as this would remove the normal deflection from the slab above.

Reshores should be located in the same position on each floor so that they will be continuous from floor to floor. Reshores must not be placed where they could cause tensile stresses where there is no reinforcement. Doing so may cause serious cracking of the concrete.

Slabs should not be reshored until adjacent beams, if any, have been reshored. Any slab with a clear span of more than 10 feet may require reshoring. Proper location of reshores should be designed in advance with the approval of the engineer-architect.

SUMMARY OF CHARACTERISTICS

Backshoring	Reshoring
• Strip small areas.	• Strip entire bay.
• Do not let slab deflect.	• Allow slab to deflect.
• Install backshores before any further stripping occurs.	• Install reshores without removing deflection.
• Slab does not carry its own weight.	• Slab carries its own weight.
• Backshores have an initial load.	• Reshores have no initial load.

For flat slabs, reshores should be placed along the intersection line of the column strip and middle strip in both directions. This should be done for each bay as it is stripped before removing forms from adjacent bays.

Reshores should not be removed until the slab or beam has sufficient strength to support all loads that will be applied to it.

With backshoring, so long as the first level of shores remains in place in contact with grade, each tier of shores must carry the weight of all concrete and construction loads above it. This may be the weight of several floors.

When reshoring remains in place at grade level, accumulated shore loads are less because each slab has been permitted to carry its own weight before the reshores were installed.

When using reshores, stripping of the formwork is more economical, since all formwork material can be removed at the same time. Reshoring generally requires fewer levels, thus freeing more area for other trades.

When using backshoring, stripping of the formwork can be done at an earlier age because large areas of concrete are not required to carry their own weight.

In North America, a majority of supported slabs use reshores rather than backshores.

Interested readers are referred to Jacob Grossman's article, "Two-day Construction Cycle for High-rise Structures Based on Use of Preshores," in the March 1986 issue of *Concrete Construction.* Preshoring, a variation of the backshoring method, is a system that has been used successfully in New York City for many years.

Stable floor forms without staking

Q. *We are consulting on a freezer floor job. The floor is to be $6^1/_2$ inches thick, with welded wire fabric, placed on a 6-inch layer of polystyrene over a vapor barrier on a sand base. The floor is large enough to require a column line down the middle, so the floor contractor plans to use a bulkhead and key joint along the centerline. The architect-engineer does not want the concrete floor contractor to use any stakes for the forms. How can the contractor achieve adequate flatness (the new designation of F 35 is specified) without having well-anchored, stable forms at the center?*

A. Some contractors are able to achieve this degree of flatness while using wet screeds (piles of fresh concrete set to predetermined levels), but if anyone tries to use wet screeds in an application like yours they will have to devise some other way than stakes to establish the levels of the tops of the wet screeds.

A good solution is to use screed rails, a development described in the article "Precast Leave-in-Place Screed Rails for Floor Construction," *Concrete Construction,* May 1984, page 483. The rails are made of reinforced concrete and serve as guides for screeding. Their design provides for interlocking of adjacent strips of floor slab and for use of tie bars across the joint if desired. Rails are set on pads of concrete and leveled with a surveyor's transit. Experience seems to show that they can be installed faster than ordinary floor forms and that they are stable during floor construction operations.

How to set forms for insulated slab

Q. *We're going to be placing a 6-inch-thick concrete slab in a cold storage warehouse. The concrete has to be placed over 12 inches of expanded polystyrene insulation board. The insulation comes in 4x8-foot sheets that are 6 inches thick, so there will be two layers. A vapor barrier goes on top of the insulation and we place the concrete on top of that. We have several questions:*

- *Will the weight of the concrete compress the insulation board?*
- *If the specs permit $^3/_{16}$ inch deviation from flatness in 10 feet, what's the best way to set our side forms so we can meet the spec? We're thinking of using 2x6-inch lumber attached to 2x8-foot sheets of plywood, then screwing plywood side forms to the 2x6s. We'd weigh down the plywood with concrete blocks or sandbags to keep the forms from moving around on the insulation when the vibrating screed is running.*
- *Is there anything else we should be worrying about?*

A. We'll answer the easy question first. A 6-inch-thick slab of concrete will put about 75 pounds per square foot of pressure on the insulation board. Regular blue board insulation has a crushing strength of 3,600 pounds per square foot, so crushing isn't a problem.

There are several possible ways to control the floor elevation other than the plywood form system you're considering, but we don't know which will work best. Some other options are using leave-in-place concrete screed rails that are set on concrete pads, or using adjustable pipe-screed supports. Can any of our readers suggest a better way to place a floor slab on rigid insulation board and meet the $^3/_{16}$ inch in 10 feet tolerance?

The only other caution is to be prepared for prolonged bleeding. With a vapor barrier beneath the concrete, excess water has to bleed out. If conditions are such that the surface dries quickly, you could have problems with blistering.

More information on forming insulated slabs

The Problem Clinic for July 1989, page 662, contains a question about placing a concrete slab over insulation board. In the answer, we incorrectly used the terms expanded polystyrene and "blue board" insulation interchangeably. The following letter explains the difference between expanded and extruded polystyrene and also gives more helpful information on construction methods.

I have a number of comments on both the July 1989 Problem Clinic question and your publication's answer. First, the question specified the insulation would be expanded polystyrene; however, your answer states "blue board insulation." In the cold storage construction industry, blue board most times equates to one of the DOW chemical extruded polystyrene insulation boards. Expanded polystyrene and extruded polystyrene insulations are two different materials, particularly when it comes to compressive strength.

Your answer states a compressive strength of 3,600 pounds per square foot (psf) which equals 25 psi. This would be the compressive strength of DOW SM extruded polystyrene. The compressive strength for the most commonly used expanded polystyrene is in the range of 10 to 14 psi which equals 1,440 to 2,016 psf. It should be noted that the compressive strength values listed are at 10% deformation when tested in accordance with ASTM D 1621.

For a cold storage warehouse, expanded polystyrene is not a good choice because of its low compressive strength. A proper slab design analysis must be performed whatever the type of insulation. But it is a fact that, given the same loading, the slab on expanded polystyrene insulation will be thicker than a slab on extruded polystyrene. The modulus of subgrade reaction for the expanded polystyrene is smaller, thus the slab must be thicker.

The idea of using a 2x6 attached to plywood is a good method for constructing slab forms and we have had years of experience with this detail. One trick that we have seen frequently is to tie the forms together with metal strap banding so they don't spread when the concrete is placed. The straps are attached to the plywood on the forms and the strap is run across the top of the insulation. It remains under the slab after placement, but it is inexpensive and it works.

A normal cold storage floor, in particular a freezer, usually has a base slab below the insulation and most times this slab encases an underfloor heating grid. This base slab should be placed to a good tolerance to achieve a good top slab tolerance. Also, the manufacturing tolerance for extruded polystyrene insulation is fairly good and consistent.

Thickness tolerances, however, are not as good for expanded polystyrene insulation. Thus if the extruded polystyrene insulation is installed properly, the top of the form system should be good. Ballasting the plywood aids in keeping the form set and, if required, shimming could be done under the plywood.

There again compressive strength of insulation comes into play. The expanded polystyrene insulation is more prone to damage if the forms are subjected to highly concentrated forces. Using this form detail, we have had cold storage floors placed with exceptional F numbers.

The question referred to the vapor barrier being on top of the insulation. The proper location for the vapor barrier is on the warm side of the insulation, which in the case of a cold room storage building floor is normally below the floor insulation. The questioner may be referring to what is known as a slip sheet. It is normally a thin sheet of polyethylene that is placed, but not sealed, on top of the insulation. This is used to keep any bleed water from migrating into the joints of the insulation and also to act as a bondbreaker. The bleed water must escape through the surface of the concrete and finishers must be careful not to seal or finish trowel the surface too soon.

Also be careful with the support of reinforcement and the control joint dowel bar baskets that are being used. Chairs for reinforcement support should have plates so the legs don't poke down into the insulation. Baskets should be used with continuous runners bearing on the insulation.

Mixing, Placing Compacting

3

Concrete

Prepare for the unpredictable

Q. *Has anyone ever written up a list of things that a contractor can readily do to prepare for some of the problems that often come up unexpectedly? We've had trucks unable to get close enough to discharge the load, we've had forms deflect and require extra concrete, we've had fresh slabs get caught in the rain, and we probably have some other kinds of unhappy experiences still in store for us. Maybe you can tell us what are some of the surprises still in store for us and how to avoid them.*

A. We suggest you start your own checklist of things you know you might do to be ready in case any of the above conditions occur. Then you can add other items that would prepare you to handle any surprises that you've heard other contractors have had. Here's a partial collection to start you off:

- For slabs on grade be sure the side forms are in place, secure, and checked for grade. Compact the subgrade and measure the depth of the slab from the top of the form to the subgrade surface by extending a straightedge from one side form to the other. Then if you allow a slight amount of extra concrete you won't be surprised by having to order more.
- Be sure the reinforcement is in place and clean.
- Have the expansion joint material available on the site.
- In hot, dry weather moisten the subgrade and forms.
- Be sure the screed, straightedge, and finishing tools are out and accessible.
- Even if it doesn't look like rain, have plenty of polyethylene on hand to cover one day's pour.
- Provide a clear path for the mixer truck, free of obstacles and debris, and be sure it's capable of supporting the heavy load of the truck and concrete.
- If trucks will be following one another, provide room for a truck to back out and leave while one or more others are waiting.
- If accessibility to forms is limited, confer with the concrete producer about supplying longer-than-usual chutes or some other adequate means of getting the concrete to the forms.
- Make a final check to see that the forms are well braced so they will stay in alignment, won't require extra concrete, and will be safe.
- Be sure the curing material is on hand in an adequate amount.

Rebar cover in grade beams

Q. *I have run into a new twist on an old routine detail in our area near the seacoast. The local code, the American Concrete Institute (ACI), and most construction plans all require a minimum of 3 inches of concrete cover over rebar in grade beams poured in earth trenches and not formed. I am inspecting the work of a contractor who is arguing that the engineered fill will settle an inch a year but that the beams themselves will not settle because they are supported by piles. His theory is that there will soon be a separation between earth and concrete; consequently 1½ inches of concrete cover is all that is necessary. The building is only about 200 feet from the waterfront. That means that the beam soffit, at 4 feet above mean sea level, is in a corrosive environment. I say the contractor is wrong and 3 inches minimum cover is required.*

A. Our inquiry to the ACI reveals that the 3-inch cover requirement is intended to compensate for the possibility of low-quality concrete in this kind of construction. When concrete is placed directly on soil there is a greater likelihood that the soil will absorb water from the mix and that the concrete in direct contact with the earth won't be of as high a quality as the rest of the structural element. The fact that the earth may later settle away from the underside of the grade beam has little bearing on the requirement for a 3-inch cover. We note that Section 7.7.1 of the ACI Building Code, which requires a minimum cover of 3 inches, indicates that this is for "concrete cast against *and permanently exposed to* earth" (italics ours). Perhaps that's the contractor's basis for his reasoning.

ACI tells us that code interpretations can be made by the ACI committee that wrote the code but that this is a lengthy and expensive process. But 3 inches of concrete cover is required.

Bar ties don't affect structural strength

Q. *Is the strength of a reinforced concrete slab or beam affected by the spacing of rebar ties?*

A. Rebars are tied only to maintain bar positions during work done by other trades and during concrete placing. Tying adds nothing to the strength of the finished structure. In most cases, a tie at every fourth or fifth intersection is all that's needed. For preassembled mats, however, tie enough bar intersections to make the mats rigid enough for handling. When snap ties are used, tie every bar intersection around the perimeter and alternate intersections within the interior of the mat as shown in the figure.

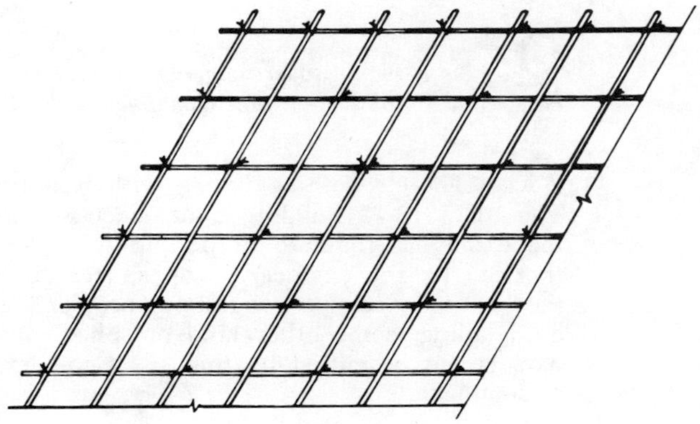

More information on bar placing, splicing, and tying is given in the Concrete Reinforcing Steel Institute (CRSI) book *Placing Reinforcing Bars.* The 193-page book is available from CRSI, 933 North Plum Grove Road, Schaumburg, Illinois 60195; telephone 708-490-1700.

Placing column dowels in footings

Q. *Is it acceptable to work the column dowels down into the fresh concrete of a footing rather than place the concrete around supported dowels?*

A. The best advice we know on this subject comes from the CRSI *Placing Reinforcing Bars,* Fourth Edition (1981), Pages 11-5 to 11-7: "Dowels must be placed accurately not only with respect to the location of the column bars above, but also vertically to assure that they project the specified distance above the footing. . . .

"The best accepted practice is to use a template made of wooden boards with holes drilled in it; or with a tie or a single turn of spiral nailed to the guide boards. . .

"It is not considered good practice to drive or push dowels into position in [fresh] concrete. They are difficult to hold in proper alignment and corrections in dowel position after the concrete has set can be costly."

Tying column cages

Q. *We've been trying to think of a way to standardize and simplify the job of tying up reinforcement cages for columns 18 inches square. There will be a large number of columns, and we would like to develop an assembly system that minimizes the work.*

A. You might investigate the use of welded wire mesh. This is widely available in flat sheets. In one direction the sheets have the needed amount of steel area and the right spacing for ties. In the other direction they have just enough wire to stiffen the cage. It is said that, when the sheets are bent on a job bender to form cages, it saves about 75 percent of the labor that might otherwise be used for tying up rebars.

This system is reported to work well for lightly reinforced columns. In heavy columns welded wire can be used as cages in combination with vertical rebars.

A little rust on rebars

Q. *In new construction should we remove the rust from rebars to be sure we get good bond to the concrete?*

A. Laboratory tests have shown that a small amount of rust actually improves the bond because it increases the normal roughness of the surface. According to the Concrete Reinforcing Steel Institute (see their book *Placing Reinforcing Bars*), bars with rust are considered acceptable without cleaning or brushing, provided that the minimum dimensions, including height of deformations, and weight of a cleaned sample are within ASTM specification limits.

Splices for reinforcing bars

Q. *I see many advertisements for sleeves and the mechanical connections for reinforcing bars, and I know they are sometimes welded. What's wrong with an ordinary lap splice?*

A. There's nothing wrong with it if there is room for it and it is done properly. The ACI 318-77 Building Code, Section 12.15.1 says "Splices of reinforcement shall be made only as required or permitted on design drawings, or in specifications or as authorized by the engineer." Lap splices are usually the most economical method, but if they cause excessive congestion, problems in placing concrete, or problems of design or detailing, some other kind of splice must be used. Lap splicing of #14 and #18 bars is not permitted by the ACI Code, except in compression to smaller size dowels at footings.

Cleaning mortar from galvanized steel

Q. *We have the job of cleaning mortar splashes over a lot of nearby galvanized hardware on stadium seats. We know that sulfates attack cement products. If we were to apply some kind of sulfate solution to the splashes would it remove them? If so, what kind of sulfate solution would you recommend?*

A. Probably Glauber's salt (sodium sulfate, $Na_2SO_410H_20$) would be the best because it would not be aggressive to zinc and is soluble up to about 9 parts of salt in 10 parts by weight of water. You might try as concentrated a solution as convenient. It would be desirable to apply it in warm or hot weather and to apply a curing compound to keep the mortar from drying out. We don't know how fast it would work on your mortar splashes, however.

Note: In the followup to this inquirer *Concrete Construction* learned that if there were any results from the sodium sulfate solution they were not fast enough to be practical. The contractor also tried penetrating oils without success. An attempt to remove by heating melted some of the zinc. Wire brushing was ineffective. Waterblasting at 3000-psi pressure did not do any good. The contractor finally removed the mortar by sandblasting, but the metal then had to be regalvanized.

Wet feet

Q. *I sometimes run into problems with water and mud while pouring footers and would like some guidelines. Can you pour concrete directly into water? Does the concrete just displace the water? How do contractors who build bridge piers set their forms? At the time the concrete is poured is there water in the forms? What effect does the water have on the concrete?*

A. The methods used for placing bridge piers require more equipment and preparation that may be justified on most foundation jobs where the amount of water involved is considerably smaller. On big projects where concrete must be placed under water the concrete is placed through a pipe of large diameter (often 10 or 12 inches) which extends below the water to the point of placement. The process is described in "Tremie Concrete," *Concrete Construction*, February 1986, page 123. Essentially the concrete does simply displace the water, but only if the work is done properly. The precautions necessary to keep the concrete from mixing with the water to any significant extent are given in Chapter 8 of ACI 304, "Recommended Practice for Measuring, Mixing, Transporting, and Placing Concrete."

There are several methods of setting forms in water for bridge piers. One is to drive piles to bedrock with their tops just below the river bottom. A coffer is built that reaches from above the water line to a point well below the tops of the piles. Within this coffer concrete is placed under water by tremie methods to produce a footing of adequate thickness that seals the bottom. The coffer is then pumped out and the pier constructed on the footing. This and other methods are described in M. K. Hurd's *Formwork for Concrete*, Fourth Edition, American Concrete Institute Publication SP-4, pages 9-54 to 9-56.

The purpose of tremie methods is to keep the concrete from becoming mixed with any significant amount of the water under which it is being placed, since any additional water would increase the water-cement ratio and weaken the concrete. If concrete were mixed with water that is muddy, the contamination by mud would weaken the concrete further.

American Concrete Institute Committee 332, Residential Concrete, has published a recommendation about water in footings: "Pools of rainwater that have collected in footing forms must be pumped out and all water that has collected in forms or on grade should be removed before placing concrete. It is not always possible to get the surface completely dry, particularly where the water table is high. If so, the concrete should be placed in a manner that displaces the water without mixing it into the concrete."

Getting uniform surface on high lift

Q. *Please give your recommendations for placing and vibrating architectural concrete in a 14-foot-high wall to be cast in a single placement. We are concerned about our ability to vibrate effectively enough to get a surface that is the same from bottom to top.*

A. A method that is reported to be effective is to provide access doors at several levels in the back face, which will not be exposed to view. These doors serve two purposes. They are entries through which concrete can be placed without dropping it long distances and they are ports through which vibrators can be inserted and operated. Concrete is placed in small batches as near as possible to its final location. Normal good vibrator practice is used to consolidate the concrete as uniformly as possible. The openings at each level are closed as soon as they are no longer needed and the openings are promptly sealed against leakage. The back face will be marred by the outlines of the openings, but presumably that face will not be visible.

Prestress loss

Q. *What do people mean by "loss of prestress"? Does this mean that the prestressing force is all going to be lost eventually, and if so, how long can you expect prestressed concrete to last, or to be safe? Or does it mean prestressed concrete isn't really dependable because it may lose some of its prestress? Does anybody know how to prevent this loss?*

A. The industry pretty well understands how both concrete and steel behave, and how concrete, steel and prestressing forces interact with one another. This makes it possible to predict approximately how much prestress will be lost in a given kind of service. In designing prestressed concrete the designer takes all of this into account—the slipping of the anchorage; the frictional loss from curvature in the tendons; and the losses caused by elastic shortening, drying shrinkage, and creep of the concrete. The result is a concrete that has the required strength even *after* loss of prestress.

Finish for overhang on earth-sheltered house

Q. *The bottom surface of the overhang on an earth-sheltered house we've built is unattractive in appearance and we must provide some kind of pleasing finish. Either rubbing or machine finishing would be very difficult to do overhead. Sandblasting wouldn't be good because there isn't enough coarse aggregate in the mix.*

A. Probably the best solution is to apply a latex-modified cement paint. There are a number of manufacturers of such products and you should be able to obtain one from your local distributor of concrete specialty materials. Otherwise check the yellow pages of our December Reference Guide, under "Waterproofing and Dampproofing Coatings."

Crushed aggregate subbase not always needed

Q. *What would happen if I omitted the crushed aggregate subbase under a warehouse floor? The soil is firm and dry, and in a well-drained location.*

A. Some authorities do not believe that special crushed aggregate subbases are needed under all slabs on grade. In many cases a slab can be cast directly on undisturbed soil or fill that has been properly compacted if there are no water problems.

The concrete mixture is very important, however. Many slabs are constructed in warehouse and industrial buildings with 3000-psi concrete but such concrete doesn't have sufficient strength or wear resistance for such service. Instead, 4000-psi or even 4500-psi concrete should be used.

Let's say it would cost about $2 a cubic yard more for 4000-psi concrete than for 3000-psi concrete. For a 5- or 6-inch slab, that would be a difference of about 4 cents a square foot. It is likely that one could save more than 4 cents a square foot by not using a crushed aggregate subbase.

Pumping

Stone sand and pumpability

Q. *We do concrete work in an area where natural sand is scarce, so most of the concrete contains manufactured or stone sand. We have a pumped concrete job coming up and I've heard that it's hard to pump concrete made with stone sand. Is this true, and if so, what can be done to improve the pumpability?*

A. The effect of sand on pumpability is related both to particle shape and to the amounts of finer sand particles. In some areas good pumpability has been achieved by using unwashed stone sands that contain higher than normal amounts of material that passes the Number 200 mesh screen. Specifications may permit the use of sands with as much as 7 percent minus-200 material. Care must be exercised in using sands with a high fines content though. Excessive amounts of fines increase the amount of water needed to get the desired slump and may also make a concrete sticky and hard to finish.

Waste the first pumped concrete

Q. *In several locations on the walls of an office building we have lighter colored concrete than elsewhere. We've determined that it has always been at locations where we began the day's work. The concrete was placed by pump. Did that have anything to do with it?*

A. Very likely. If the first half cubic yard or so of concrete through the pumpline wasn't wasted, it could account for the trouble. That first concrete loses some of its cement paste and mortar in the process of coating the pipeline and probably changes the water-cement ratio slightly as well. Both effects change the color. Either waste the first half cubic yard of concrete or prepare ahead of time to use it for some other and noncritical application.

Change to drier coarse aggregate in pumping mix

Q. *On a good-sized pumping job there was serious trouble with fouling of the pump. Everything had been going well with load after load of 4000-psi, 4-inch-slump, air-entrained concrete when suddenly every load became virtually unpumpable. It subsequently turned out that the plant had used up its stock of the moist coarse aggregate being used and had started to draw from another pile of aggregate, which happened to be dry. Is this the likely explanation?*

A. Such a change can cause the trouble you describe. When the dry coarse aggregate was batched using the same batch weights as for moist aggregate it required more rock than normal to provide the same weight as the moist aggregate. If there had been, for example, 1750 pounds of coarse aggregate, and the total change in aggregate moisture content was 2 percent, it would mean that about 35 pounds more dry aggregate was batched. This would make the mix more harsh. At the same time, the drier aggregate would reduce mixing water content by about 35 pounds, partially drying out the mix. The two effects would reduce pumpability.

It pays for the pumping contractor to know something about mix design, as discussed in the 2-page article in the September 1982 issue, page 721.

Minimizing wear on pump hose line

Q. *For our pumping operations we've been using hose that is supposed to have a wear-resistant cover. Yet we've seen the hose wear out in places, sometimes after a fairly short period. Should we be looking for a different type of hose or is there something we can do to make it last longer?*

A. Wear-resistant hose doesn't resist everything. Some things to avoid are laying it against sharp objects or letting it hang over sharp corners. Such conditions may produce hose failures quickly. Kinks in the hose should be avoided; they can burst the hose after only a few cubic yards have passed through it. It is also a good idea to lay hose on plywood or other smooth material rather than directly over rebars. Then check the hose for wear daily. If a weak spot is noticed, take the hose out of service. You will avoid having it burst in the middle of a job. The hose supplier should be able to inspect and repair the hose if necessary.

Pump to reach center of building

Q. *Our company has been awarded the floor contract for a hospital. We will have to reach out as much as 200 feet and place 50 cubic yards per hour to make slabs up to 40 feet wide at heights up to 75 feet. Our cranes will not reach far enough to place concrete in the middle of the building. What alternative would you suggest that we could use without any backup method?*

A. A small pump could move concrete at this rate for the distances described. A dual 4-inch line system could be used to place the 40-foot-wide slabs.

Clean the mixer truck

Q. *The 4-inch line from a pump plugged up recently. It turned out that there were some hard lumps of concrete in the mix and a number of these had collected together to plug the line. We had to clean out the pump hopper and discard part of the load. Could these lumps have formed enroute to the job?*

A. This problem has been experienced when the truck mixer has not been properly cleaned after one or more previous loads. When something happens during mixing to break the hardened concrete loose from the vanes the lumps get distributed through the fresh mix. Sometimes when a large piece breaks loose it helps break additional pieces loose. Careless cleaning of the truck mixer can lead to trouble.

Lumps beyond driver's control

Q. *A mixer truck driver, David Bryant, did not agree with our answer to the Problem Clinic "Clean the mixer truck." He argues that lumps in concrete are not the driver's fault.*

A. The following are only the most obvious reasons for lumps in concrete:

- Unclean trucks. If this is the case, in a very short time the truck would become so dirty inside that it would be obvious.
- Improper loading of transit mixer. This can happen because of:
  1. leaving the water running while loading (produces severe chunking)
  2. loading the mixer slower or faster than it was designed to be loaded
  3. overloading the mixer (on extremely small loads of 1 cubic yard or less)
  4. too short a haul between the plant and the jobsite (not just too short a mixing time, since road vibrations shake lumps loose from the blades as nothing else does)
  5. lumps of cement in the cement silo

The original question asked was whether lumps could develop en route to the job. Your negative answer was totally inaccurate. If at the time of loading, unmixed dry cement on the top of the load absorbed just enough moisture to begin the chemical reaction, 30 minutes of transit time followed by a 45-minute delay on the job (both of which are common) would be more than enough time for lumps to develop and become quite hard.

Editor's Note: We appreciate David Bryant's thorough discussion and his catalog of possible causes of lumps and we are willing to concede that failure to clean the truck may be one of the least likely causes. The cause described in the Problem Clinic item cited is one that has, however, actually been experienced. If the lumps are initially harder than lumps could become in the first two or three hours, and if they are not made up of virtually pure cement from the silo, they have probably been caused by not cleaning the mixer thoroughly.

Ready mixed concrete

Hold the deliveries

Q. *We are placing a floor today and the ready-mixed concrete trucks are arriving on the job faster than we can get the concrete discharged and placed. How long can we allow the concrete the remain in the truck before discharging it?*

A. This question is covered in ASTM C 94, "Standard Specification for Ready Mixed Concrete." Section 10.7 of the standard notes: "Discharge of the concrete shall be completed within $1^1/_2$ hours, or before the drum has revolved 300 revolutions, whichever comes first, after the introduction of the mixing water to the cement and aggregates or the introduction of the cement to the aggregates. These limitations may be waived by the purchaser if the concrete is of such slump after the $1^1/_2$-hour time or 300-revolution limit has been reached that it can be placed without the addition of water to the batch. In hot weather, or under conditions contributing to quick stiffening of the concrete, a time less than $1^1/_2$ hours may be specified by the purchaser."

Concrete rejected because it wasn't discharged soon enough

Q. *Our country road inspector recently sent whole or partial loads of concrete back to the plant because our curb and gutter crew didn't get the concrete out of the truck within one hour after load time at the plant. What are the time guidelines here and is there any flexibility?*

A. The ASTM Standard Specification for Ready Mixed Concrete (ASTM C 94) requires discharge of the concrete to be completed within $1^1/_2$ hours, or before the drum has turned 300 revolutions after the cement comes in contact with water. However, the specification also allows these limitations to be waived by the purchaser if the slump is such that the concrete can still be placed without adding more water. In hot weather, the purchaser may specify a time less than 1½ hours. Judgement is required in determining whether the concrete can be placed properly after the time limits have been exceeded.

Killing the set of concrete in disabled transit mixer

Q. *If a truck mixer will be broken down for an extended period, but we can still turn the mixer drum, what can be added to the concrete to completely destroy its set so as to prevent the concrete from hardening in the drum?*

A. The old standby for destroying the set of concrete is ordinary sugar. Many concrete mixer trucks carry a bag of sugar for just such emergencies. For most circumstances 15 pounds of sugar ought to be enough to kill the set of a 10-cubic-yard batch if it can be thoroughly mixed in. It would be best to completely dissolve the sugar before adding it to the mixer. Some companies prefer to carry excess retarder on the truck and add whatever amount they feel sure will be effective. The concrete is wasted but the mixer drum remains usable.

Saving the mix in disabled transit mixer

Q. *Occasionally one of our mixer trucks breaks down before reaching its destination. Our problem is what to do with the load of fresh central-mixed concrete the mixer is carrying. Frequently the truck will be down for a period of one to three hours. Can commercial retarders or other products be added to the concrete to keep the mix alive and still of good quality until the repairs are completed and the truck is underway again? (Assume that we can still turn the drum while the mixer truck is broken down.)*

A. Saving the concrete, yet not prolonging the set so much that it will disrupt job operations, is something of a neat trick. The time set could be extended by adding a retarding admixture. But the amount to add would depend on whether or not there is already a chemical admixture in the batch. It would also depend on how long it has been since the concrete was originally mixed (because that will determine to what extent the early hydration reactions have proceeded). Furthermore, the time of year will have an effect, because hydration is faster in hot weather than in cold. You might run a test mix to which, after some selected period of time, you add about half a normal dose of retarder, then determine what the retarder has finally done to the hardening time and the 28-day strength. Then, maintaining the same job conditions, you could make the same measurements with small or larger dosages, depending on what seems to be indicated by the results of the first mix.

You must bear in mind that the effect of the admixture will change with any of the changes in conditions already discussed, but also that it is likely to change with any change in cement brand or change in amount of pozzolan. If the subject is of sufficient importance to your operations you could try to determine the proper amounts for each of the material and environmental conditions anticipated and establish a chart for future guidance.

Saving the drum of disabled transit mixer

Q. *If a truck mixer remains broken down for an extended period of time and the drum cannot be rotated, what is the best thing to do?*

A. We know of no good way to handle the situation if the drum cannot be rotated. The main concern is how to save the drum. Most companies make it their basic strategy to do what they can to get the drum turning again. Drivers should be instructed that if the drum will not turn they must notify the plant immediately so that all necessary help can be recruited quickly.

For the worst case—that the truck is overturned—vigilant ready-mix companies try to be prepared by keeping track of where they can promptly mobilize a crane or other lifting rig, and they become familiar enough with each truck so that they understand the proper lifting points. In the process of lifting, some concrete is likely to spill out. It may be desirable to try to empty the drum as much as possible before lifting by getting a local fire department to flush it with fire hoses, but this depends on whether both concrete and water can be made to drain from the opening or if the access plate is at a low point and can be removed for better drainage.

If the truck is upright and the drum is hydraulically driven, a simple solution is to bring up another truck and connect to its hydraulic system. This works if it is the hydraulic pump or line and not the hydraulic motor on the disabled truck that has failed. If there is a failed hydraulic motor, mechanics should be brought to the scene promptly with a replacement motor that has been kept in readiness for just such an emergency.

One way to get a mechanically driven mixer drum turning is to put a rubber belt around it and the drum of another mixer truck which has been drawn alongside. This belt might be a length of conveyor belt.

If the drum can only be turned temporarily, the bolted-on plate, or hatch, on the side of the drum should be removed and the drum turned until the resulting opening points downward. Meantime a local fire department should have been called to flush out the drum with fire hoses. Cleaning up the gutter or site is easier than removing hardened concrete from the drum.

It has been suggested that one could try to inject carbon dioxide from a gas cylinder, using a long probe reaching to the bottom, and moving the probe from time to time to reach all portions of the mix. The purpose would be to produce noncementitious products, the way the carbon dioxide from unvented heaters produces soft, low-quality floor surfaces. However, we feel that because of the many cementitious compounds present in the mix, all in different stages of hydration, the result would be quite unpredictable; one possible result could be *accelerated* hardening.

Similarly, cooling with liquid nitrogen also seems impractical because of the tremendous quantities that would be required to lower the temperature enough to delay hardening for any substantial period.

Finally, all efforts to turn the drum and discharge the load sometimes do fail and the truck eventually gets back to the yard filled with hardened concrete. In such a case the access panel can be removed from the drum and the mass broken up with a hydraulic splitter. One truck containing 5½ cubic yards was cleaned out in 5 days by this operation (see "Splitter Breaks Up a Hard Problem," *Concrete Construction,* February 1982, page 208). Alternatively, the concrete might be broken up with a chemical splitting agent (as described in "Nonexplosive Demolition of Concrete and Rock," April 1982 issue, page 366).

Vibration

Internal vibration of flatwork

Q. *We hear recommendations both for and against using internal vibrators in constructing floors and other flatwork. What's best?*

A. One of the reasons for cautioning against the use of internal vibrators is concern over segregation and the possibility that this may happen if workers use internal vibrators to move piles of concrete into position. It can also happen if they use them to vibrate high slump concrete that doesn't require vibration for placement. However, if there are rebars in the floor, internal vibrators are useful in helping consolidate the concrete against the steel. Also, in any slab that is 8 inches thick or more, internal vibrators should always be used so that the full thickness becomes uniformly consolidated.

Constructing near blast sites

Q. *Construction of a 5-story building about 100 yards from a working quarry is under consideration. How damaging would the quarry blasting operations be to the concrete? Should concrete be allowed to reach some minimum age before it is exposed to shocks from blasting? We want to know whether we would have to work out a careful schedule, perhaps in cooperation with the quarry operators.*

A. There have been a number of studies that show that concrete is remarkably resistant to damage from jarring and vibration at early ages and that *some* jarring and vibration may make it stronger. Nevertheless, there must be some amount of jarring that would be too much. In "Damage Effects of Pile Driving Vibrations," *Highway Research Board Record*, Number 155, 1967, John F. Wiss has discussed how to make measurements of displacement and frequency of vibrations. From these measurements a peak particle velocity can be calculated which is a guide to whether the jarring or vibrating is excessive. The allowable peak particle velocities are:

Age at which vibration occurs, hours	Permitted peak particle velocity, inches per second
Less than 10	0.1
10 to 24	4.0
More than 24	7.0

If you could make some measurements at the jobsite when blasting is being done you could use Wiss's method to calculate what the peak particle velocities are. Then you would know whether you have to take any precautions and if so, how long your concrete must be kept vibration-free.

The vibrator knows what it wants

Q. *We have recently been having trouble with vibration of concrete in wall forms. Normally we try to give good, thorough vibration with an immersion vibrator, but if we vibrate the mix very long on this job we get excessive mortar rising to the surface. The concrete comes from a supplier whom we have not purchased from previously. We have reduced the slump as much as we can, and that helps a little bit but not enough. Many people say you can't overvibrate concrete, but we seem to be overvibrating this concrete.*

A. Perhaps you have an improper mix design, with too little aggregate. You might modify the mix design by making a series of mixes through which you gradually increase the aggregate content and possibly reduce the water-cement ratio further. You could vibrate each mix of the series and choose the mix in which the vibrator doesn't cause the fines to separate.

This is an interesting point that was brought out a generation ago by an English investigator, D. A. Stewart. He found that mortar came to the surface when he vibrated a mix that had a water-cement ratio of 0.52 and a cement-to-aggregate ratio of 1:6. He continued to add fresh concrete and vibrate the mold, letting the separating mortar spill over the top until he got a mix which would not accept any more coarse aggregate. He then analyzed what was in the mold and found that the water-cement ratio had dropped to 0.49 and the cement-to-aggregate ratio had changed to 1:7.3.

He then made a new mix of the proportions that the vibrator had thus selected for him and found that the mix compacted well, with no segregation. This mix, with its lower water-cement ratio and lower cement-to-aggregate ratio, was obviously less expensive than what he started with.

It is reported that in a mix with too much coarse aggregate the coarse aggregate will segregate at the top in the same way. Hence, the vibrator can be used to establish the proportions that it can handle most efficiently to produce the densest mix.

Placing concrete on frozen subgrade

Q. *I've always been told never to pour concrete on frozen subgrade. Why is that a problem?*

A. There are several reasons for not placing concrete on a frozen subgrade. Some soils expand or heave when they freeze. When they thaw later, the ground subsides and the footing or slab loses subgrade support. This can cause the concrete to crack.

If flatwork is placed on frozen ground there's another danger. The cold concrete at the bottom takes longer to set than concrete near the surface. The surface crusts over and looks like it's ready to finish while the rest of the concrete isn't firm enough to support the finishers' weight. Concrete may bulge out underneath the power trowel, creating a wavy surface. Or the surface may crack.

Finally, there's the possibility that the cold subgrade will freeze part of the concrete in contact with it. Freezing damages green concrete.

Wind chill and the freezing of concrete

Q. *In addition to the regular air temperatures reported in winter weather forecasts we keep hearing about the wind chill as well. How does that apply to concrete? When the air temperature is above 32° but the wind chill is well below 32°, does this mean that the concrete could freeze and must be protected accordingly?*

A. The quick answer is no. The concrete will not freeze when the air temperature (ambient temperature) is above 32° no matter how frightening the wind chill is reported to be and no matter how hard the wind blows. At any ambient temperature below 32°, however, the chance of concrete freezing increases with an increase in wind velocity.

The wind chill only relates to the loss of heat caused by cold air moving over a warm surface. It compares the rate at which heat is lost from a warm object, when the colder air surrounding it is still, to the rate at which heat is lost when there is a wind, with the ambient temperature the same in both cases.

The wind chill, as reported by the weather service, is based on the effect of cold temperatures and wind velocity on the human skin from research done by the U.S. Army. If there were a wind chill for concrete, however, the principle would be similar. For example, if concrete at a temperature of 60° were exposed to an ambient temperature of 35°, and if there were no wind, the concrete would begin to cool slowly. If a wind began to blow, a wind chill factor would be created and the concrete would cool more rapidly. If the wind chill for concrete were reported as 25°, it would merely mean that the rate at which the concrete would cool at a 35° ambient temperature with the wind blowing at a specific velocity, is the same as it would be if the ambient temperature were 25° and there were no wind. In neither case can the concrete be cooled below the ambient temperature of 35°. With the wind blowing it only cools faster.

When the ambient temperature is below 32°, however, wind velocity does affect the likelihood of concrete freezing. With exposure to low temperatures, hydration is slow, and little heat is generated from within the concrete. With a below freezing ambient temperature, rapid cooling could very well bring the concrete down to the freezing temperature, with the surface being especially vulnerable. These facts were known for many years before wind chill was popularized to provide an ominous note of interest in television's winter weather news and to add to the national confusion index (NCI).

Compressive strength guide for freeze resistance

Q. *How do I resolve the following contradictory published recommendations on protecting fresh concrete against early freezing? Your reprint collection,* Curing Concrete: Methods and Materials, *says on page 11, "Protection should not be discontinued before a minimum compressive strength of 1500 psi is reached; it is assumed that by this time capillary pores in the concrete will have emptied sufficiently to avoid damage if the concrete freezes." In* *three other places I find similar explanations, but the value given is 500 psi. These three references are:*

- *ACI 306R-78, "Cold Weather Concreting," Section 1.10.1*
- Concrete Construction, *"What To Do With Concrete in Cold Weather," September 1984, page 789*
- Concrete Construction, *Problem Clinic, "How Accelerators Protect Against Freezing," April 1985, page 372*

Which recommendation is right?

A. The recommendation accepted today is 500 psi. The article in the reprint collection that gave 1500 psi was originally published in the September 1964 issue of *Concrete Construction*, page 270. This article showed data originally published in the article by H. J. Gilkey, "Curing Structural Concrete," *Journal of the American Concrete Institute,* May 1952, page 711. In that article the author said "It is assumed that protection against freezing may be discontinued when the concrete has attained a compressive strength of 1500 psi. By this time the capillaries are believed to be sufficiently emptied to obviate freezing damage." By 1956, however, ACI 604-56, "Recommended Practice on Winter Concreting," was setting this value at 500 psi, the same value still given in the three later references you cite. ACI 306R-78, Section 1.10.1 now reads "Prevent damage to concrete from freezing at early ages. The degree of saturation of newly placed concrete, which has no access to external sources of water, will be reduced as the concrete hardens and water is combined in the hydration process. Under such conditions, the time at which the degree of saturation becomes reduced below that level which would cause damage by freezing corresponds roughly with the time at which the concrete attains a compressive strength of 500 psi. . . . At temperatures of 50 [degrees] F. . . . most well proportioned concrete will reach this strength during the second day."

Rainy weather

Q. *In our part of the country we don't have much trouble with rain. However, we have a project coming up that must proceed on schedule and we are concerned about what might happen if we should ever have to shut down on account of rain. Are there any standard practices for proceeding with concreting in rainy weather?*

A. Placing concrete in the rain should ordinarily be avoided, but if concreting is absolutely necessary all precautions should be taken to keep additional water from the fresh concrete. Following are some measures recommended by Robert F. Adams in *Concrete Construction Handbook*, edited by Joseph J. Waddell:

- Cover the working area with tarpaulins or other protection.
- Use lower-slump concrete than otherwise (since evaporation practically stops during a rain).
- Before placing concrete on footings or on construction joints, dry up any puddles of water already standing on them.
- Provide a location for water to drain off the top surface, and try to maintain a slight slope on the fresh concrete surface if possible so that bleed water will move toward the drainage point.
- Avoid working the surface of the fresh concrete.
- During thunderstorms of short duration, stop placing concrete until the storm passes but meanwhile cover the concrete with tarpaulins, plastic sheets, or other effective protection.
- If rain is so heavy that puddles cannot be removed and the concrete is being washed, it may be advisable to suspend concrete placing. If so, bulkheads, dowels, headers or steps should be put in so that the concrete already placed is left in the best condition possible for continuing the work at a later time. This process creates a construction joint and when work is continued the cleanup practices normally recommended at construction joints should be followed.

Temperature and humidity effects on productivity

Q. *We're bidding a concrete forming job that will be done during the summer in the southeastern United States. In estimating the time required to finish the job, I'm concerned about the effects of heat and humidity on worker productivity. Are there any guidelines for judging how much work will be slowed by hot weather?*

A. Jim Adrian's book, *Construction Productivity Improvement,* has a whole chapter on productivity and the environment. Two tables from the chapter are shown here. The top table shows the effect of temperature and humidity on productivity. If the effective temperature increases from 70° F to 100° F at 60% relative humidity, productivity drops by 20%. In cold weather, effective temperature depends on wind speed as well as the actual thermometer reading. Use the other table for determining effective temperature during cold weather. The tables were developed by the National Electrical Contractors' Association but are probably accurate enough for estimating weather effects on productivity for most construction tasks. Adrian's book was published in 1987 by Elsevier Science Publishing Company.

EFFECT OF TEMPERATURE AND HUMIDITY ON PRODUCTIVITY

Productivity under various environmental conditions (%)

Relative humidity (%)	−20	−10	0	10	20	30	40	50	60	70	80	90	100	110	120
90	?	56	71	82	89	93	96	98	98	96	93	84	57	0	?
80	?	57	73	84	91	95	98	100	100	98	95	87	68	15	?
70	?	59	75	86	93	97	99	100	100	99	97	90	76	50	?
60	?	60	76	87	94	98	100	100	100	100	98	93	80	57	?
50	?	61	77	88	94	98	100	100	100	100	99	94	82	60	?
40	?	62	78	88	94	98	100	100	100	100	99	94	84	63	?
30	?	62	78	88	94	98	100	100	100	100	99	93	83	62	?
20	?	62	78	88	94	98	100	100	100	100	99	93	82	61	?

Effective temperature (° F)

EFFECTIVE TEMPERATURES

Wind speed (mph)	Actual thermometer reading (° F)									
	50	40	30	20	10	0	−10	−20	−30	−40
Calm	50	40	30	20	10	0	−10	−20	−30	−40
5	48	37	27	16	6	−5	−15	−26	−36	−47
10	40	28	16	4	−9	−21	−33	−46	−58	−70
15	36	22	9	−5	−18	−36	−45	−58	−72	−85
20	32	18	4	−10	−25	−39	−53	−67	−82	−96
25	50	16	0	−15	−29	−44	−58	−74	−88	−104
30	28	13	−2	−18	−33	−48	−63	−79	−94	−109
35	27	11	−4	−20	−35	−49	−67	−82	−98	−113

What to do with a rained-on surface

Q. *We were pouring a slab for an outside loading area and had a sudden heavy rain on the last day of a two-day pour. We tried to refinish the surface with a grout but it's not holding. The rest of the slab has a smooth troweled finish, and the owner wants the whole slab to look the same. Short of replacing the slab, is there anything I can do?*

A. If you had adequate air entrainment in the concrete for durability, yes there are some possibilities. One of our most impressive exposures to this problem occurred some years ago on a small street paving project. To complete the job, a small fill-in section had to be placed around a curb radius in each of two intersections a block apart. Just after placing, both sections were rained on, but the finishers had time to refinish only one of them before the concrete had set. After one winter, the refinished surface was 100 percent scaled. The surface that had not been refinished had not scaled. It still had the same grainy surface texture that the rain had caused by washing away the surface mortar the year before.

There are two lessons here:

1. As far as surface durability is concerned, refinishing a wet rained-on surface, unless done with great care, is likely to be worse than doing nothing at all.

2. If refinishing proves unsuccessful, removing the weak surface material from the hardened concrete should leave an exposed surface that is durable. That assumes, of course, concrete of adequate quality was placed.

This lightly scarified surface resembles a broomed finish. Especially for outdoor flatwork where a non-skid surface may be preferred, scarifying can be a remedy for unsatisfactory surface finishes.

A light grinding or scarifying of the surface may work well and be the most economical way to achieve a uniform surface. There is a variety of equipment available, large machines for highway work and smaller walk-behind equipment that would be suitable for your purpose. To achieve uniformity, scarify the surface of the entire slab, both the troweled and the rained-on areas, using a walk-behind scarifier with lightweight cutters. The unsound material will be removed from the rained-on area and the entire slab will have a surface texture similar to a broomed or burlap-drag finish. It should be a reasonably uniform non-skid surface that might be preferred over a smoothly troweled surface that can be slippery when wet. You should be able to match surface textures closely enough that minimum wear and weathering will give you satisfactory uniformity.

Another possibility is grinding only the rained-on surface with a smooth cutter, producing a surface similar to terrazzo. It is as close as you can get to the smooth troweled finish, but it is not likely to be a satisfactory match.

In either case the cost should be well below the cost of replacing the slab. The equipment needed is manufactured by several companies and is available through equipment rental outlets.

Finishing, Jointing, Curing

4

Concrete

How to finish flatwork

Q. *Where can we get a good manual on concrete finishing?*

A. We'd like to tell you about two good books on the subject. One is *Slabs on Grade*, an 80-page (coat-pocket size) book in the Concrete Craftsman Series published by the American Concrete Institute, Box 19150, Detroit, Michigan 48219. The book discusses materials, mix proportions, job operations and common problems, and it has an introductory chapter on what the concrete craftsman should know about concrete.

Another book that covers the subject well is "Guide to Finishing and Related Problems," by J. C. "Skip" Yeager. This book is available from *Concrete Construction*, 426 South Westgate, Addison, Illinois 60101. This 32-page book discusses construction and finishing methods for regular, heavy-duty and ornamental flatwork surfaces; curing methods; and what to do about problems. It also discusses the finishing of vertical, architectural surfaces.

Getting rid of bleed water

Q. *What is the best way to remove excess bleed water from a slab so that finishing isn't held up?*

A. It's best to use concrete that doesn't bleed excessively. One recommended method for removing bleed water is to drag a rubber hose slowly over the surface; the concrete should be stiff enough so that only water is removed. Then before troweling, make sure that bleed water isn't continuing to collect. Another method is to use fans or blower heaters to evaporate the excess moisture. This may work if the concrete is setting slowly enough to allow time for further finishing operations to be complete after the excess water is gone.

Trowel chatter

Q. *Why do we sometimes see a series of ridges along the radius lines in the sweeps of the trowel on a floor surface? It doesn't happen often, and I haven't been able to figure out what's different when it does.*

A. Usually this means that the finisher has tilted the trowel face too steeply on the last troweling, causing it to chatter and leave these lines.

Finishing superplasticized concrete

Q. *We're generally sold on flowing concrete but we've had trouble in finishing flatwork. The mixes tend to be sticky. Finishers find that the concrete moves under their weight, it drags on the trowel, and the surface sometimes tears. Has anyone found what to do about such problems?*

A. Yes, others have noticed these problems. It helps to use more coarse aggregate or a coarser grading of the fine aggregate.

Another way is to adjust your timing. The trouble is that the concrete appears to be ready to finish before it has actually reached initial set. This happens because there is less bleed water available; the concrete has become stiff from shortage of water, not from setting. Some manufacturers suggest waiting about 20 minutes after the glossy sheen is gone before starting the finishing operations.

Exposing aggregate on treads and risers

Q. *In exposing aggregate on both the treads and risers of concrete steps, can I use a retarder on the forms?*

A. Yes, you can apply a surface retarder to the form surface to retard the concrete adjacent to the form. That should permit you to expose the aggregate on the risers after the forms are stripped. It may take some experimentation to learn how much retarder to put on the forms and how long the retarder will delay the surface set. You can also use a retarder on the treads but it may be difficult to get the same degree of retardation on the treads as on the risers. If so, you may wish to *plan* on a difference and try to get some contrast between the degree of exposure on the two kinds of surfaces.

Finishing floors by grinding

Q. *We will be building some low-rise apartments in early winter and we are concerned about the long waiting time before finishing. Would it be economical to finish by grinding?*

A. This method has proven successful and economical on a number of jobs in recent years. Essentially the method is to place, screed, and bull-float, cure for 2 to 5 days and then grind with a grinder-finisher. At this early age the grinding is relatively easy and it is said that one man can grind about 4000 square feet in a day.

Hand rubbing and weather protection

Q. *After the forms have been removed does hand rubbing the surface actually improve the weather resistance of the concrete? Are there other methods that will provide the same amount of protection?*

A. Hand rubbing is done mainly to produce uniform appearance—to fill bugholes or overcome blotchiness or unsatisfactory color.

The durability of concrete in freezing and thawing exposures is seldom a problem on vertical surfaces. An adequate air content in the concrete mix can be used to protect against freeze-thaw damage.

Other potential effects of weathering—chemical attack by atmospheric pollutants, collection of airborne dirt and dust, nonuniform collection of such contaminants, popouts from certain components or contaminants in the aggregate, staining from certain materials used in or with concrete, efflorescence, and cracking or spalling from building movements—are special problems requiring other measures. These are discussed in a series of three articles, "The Weathering of Concrete," "Designing of Concrete for Weathering" and "Constructing Concrete for Weathering," which appeared in the January, February and March 1972 issues of *Concrete Construction.*

Safety with bull floats

Q. *Recently we put in a floor very close to some power lines. When we realized that the handles of bull floats could easily come into contact with the power lines, we quickly slit some large rubber hose to slip over the handles and bound it in place with electrical tape. Then we warned the finishers to be careful not to let the handles touch the wires even though they'd been protected, but it made us nervous. Does anyone make bull floats with some kind of approved rubber-insulated handles?*

A. This is a safety matter to which contractors and finishers should keep alert. Fiberglass-reinforced plastic handles are available to meet OSHA standards for safety in such circumstances. You can use them to replace regular handles, at least when working near power lines.

A related safety matter is keeping alert to see that bull float handles don't hit trucks or get hit by them.

Concrete burns skin

Q. *A customer of ours developed severe chemical burns on his knees when placing concrete in his hog barn. These burns may eventually require skin grafts. The concrete mix he was using contained 2 percent calcium chloride. Could the calcium chloride have aggravated the burning action of the concrete?*

A. All concrete can eat away, or dissolve skin. Calcium chloride has little or no effect on this action. Concrete burns are insidious; often the victim is unaware that anything is happening until the action is far advanced. For this reason ready mixed concrete producers are beginning to issue warnings to users about the dangers of exposing skin to concrete and to describe precautions and remedial measures to take. For good suggestions on this, see the article by William G. Hime, "Let's Stamp out Concrete Burns," *Concrete Construction*, April 1982, page 361.

Flatness costs money

Q. *Some of your articles ("Can Your Floors Pass This Test?," June 1981, page 512; "The Floor Tolerance Conundrum," August 1981, page 673; and "A Time Bomb in Floor Tolerances," November 1982, page 865) are making me a little more cautious on my estimates. Realistically, tolerances of 1/8 inch in 10 feet appear to be hard to meet. Has anyone ever made a study of the relative costs of various tolerances on floor flatness?*

A. We have recently come across some data compiled by a large floor contracting firm based on their own operations only. The chart shown here, based on their data, shows how the deviation allowable under a 10-foot straightedge affects the cost of materials plus labor for a 6-inch floor. The plotted points represent deviations the company considers acceptable for various kinds of floor use, and these points show the relative costs. (Since there are a number of possible ways being used to make tolerance measurements, it should be said that these data are based on the departures from a 10-foot straightedge supported on shims of equal thickness.) The cost of an abrasion-resistant floor for foot traffic only, with a tolerance of 0.5 inch in 10 feet, was taken as the standard cost. Relative costs of abrasion-resistant floors with lower tolerances are shown along the upper curve, and they increase progressively up to 150 percent for a floor for turret truck aisle traffic. The company found costs for all floors with ordinary, nonabrasion-resistant finishes to be lower (see lower curve). Percentage differences for all floors would have been greater, the company reports, if the costs shown had been for labor only. If such a plot is representative of most contractor's costs one could readily conclude that the all-too-commonly specified tolerance of 1/8 inch in 10 feet is well beyond many owners' price range, but that such a small tolerance is not often needed anyway.

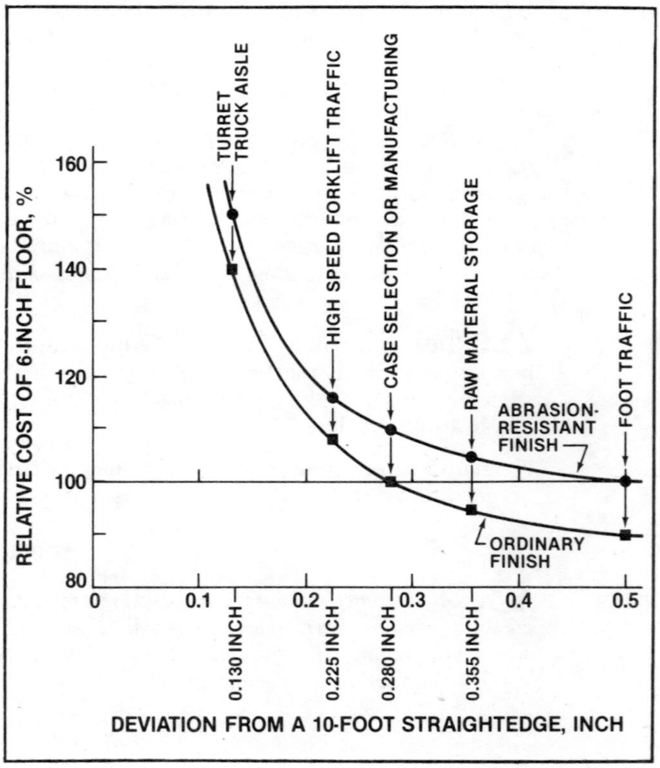

Not-so-sudden warning

Q. *You published a notice on page 372 of the March 1984 issue about an ACI Committee 117 warning to people about specifying floor flatness within 1/8 inch in 10 feet unless needed for special purposes. The committee also warned contractors not to bid on such jobs without taking into account the specialized placing and finishing techniques that are needed to construct them. Why the sudden warning?*

A. The warning is not so sudden. In the past several years the industry has become increasingly aware that specifiers have tended to specify 1/8 inch in 10 feet or something similar without considering whether such a tight tolerance is actually needed. Contractors have bid on such floors and built them. Very seldom have the completed floors been measured to see that they comply with specifications. When the floors have performed satisfactorily with respect to flatness—as they almost always have—the owners, the specifiers and the constructors have almost universally assumed that the floors met the specifications. Much evidence of recent years has shown that this is not so. When it has been discovered that a floor does not meet specifications it has frequently led to unnecessary contractual and legal disputes. Everyone should become more aware of what is really needed and what it may cost.

Floor flatness specification needed

Q. *We've been searching for published standards for the tolerances allowed in finishing residential flatwork. Are there any you know of? Specifically, are there any standards for basement floor slabs that are intended for possible future living space? We have a 3-inch drop within distances of 6 to 10 feet in a basement floor. If we apply a topping, the floor level will be above the bottom of a door.*

A. Actually, there are two main considerations: flatness and levelness. Flatness is concerned mostly with the small, shallow waves of about a foot in length. Levelness is concerned mostly with the major tilting, upward humping or downward dishing of the floor. Recently developed studies and methods of measurement are making it easier to sort these out and to express the flatness mathematically by a single number known as the F number. ACI Committee 117, Tolerances, is currently preparing a document that is expected to assign reasonable F numbers for flatness and levelness for various kinds of use. This committee document is recent and has not yet been standardized, so its values are still not officially recognized.

Nevertheless, the values you cite are very large. Based on common practice and experience they would undoubtedly be considered unacceptable by most authorities.

Shallow waves in suspended floor

Q. *This sketch (next page) shows a ripple problem with a floor finish on a large job we have just completed (8 stories totaling 200,000 square feet of floor).*

The floor fill was a lightweight material composed of 5/8-inch structural lightweight aggregate and 564 pounds cement per cubic yard. The mix was designed for pumping but was actually dumped into place with crane and buckets. Most of the time the temperatures were above 80 F, though sometimes lower. Concrete was leveled by hand screeds to high spots on the metal ducts and machine troweled. Ripples in the finish showed up to some degree on the entire area where lightweight concrete was used. On the entire first floor where the raceway was placed directly on the subslab and slightly more than 3 inches of stone concrete fill was used, the ripple either did not occur or was barely noticeable.

Have others had this trouble? What is the cause? What could have been done to correct it?

A. We do not recall encountering this problem before, though it could easily have happened to others. The conditions suggest that you are experiencing trouble from different amounts of subsidence of solids before the concrete hardens. The concrete varies from 5 1/2 inches thick to about 3/4 or 1 inch thick. There should be more bleeding, hence more subsidence, where the fresh concrete is deep. This explanation is reinforced by the fact that there was less waviness on the ground floor, where there was less difference in thickness. It is also possible that there was less bleeding in the normal weight concrete used in the first floor than in the lightweight concrete used in the suspended floors.

Two possible methods of preventing the trouble might be:

- vacuum dewatering, to remove most of the water before finishing operations begin
- constructing a two-course floor in which the topping is equal to the thickness above the ducts.

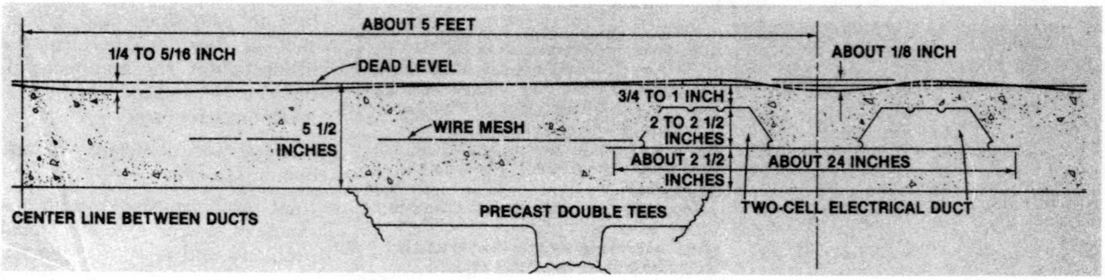

Carborundum for slip resistance

Q. *I need some information about the application of carborundum chips to walks or driveways. I have a customer who wants this done on a new home he is building.*

A. ACI 302.1R-80, "Guide for Concrete Floor and Slab Construction," Section 4.2.4.1, mentions the use of emery as a slip-resistant aggregate, and in its Sections 7.12.1 and 7.12.2 it describes how to apply them. The method is much like the method described in the article on page 285 of the March 1984 issue, "Dry Shake Hardeners." If carborundum were to be used instead of emery it would be applied in the same way.

Carborundum is harder and more abrasive than emery. We do not recall hearing of it ever being used for slip resistance—perhaps because emery is generally considered adequate and probably less expensive. Another way to achieve slip resistance is to broom the surface. After final troweling, a push broom—or even a stiff-bristle broom—is drawn over the surface. On most sidewalk or driveway jobs the troweling is omitted entirely to avoid introducing slipperiness, and then these jobs may or may not be finished by brooming.

If your customer is looking for both slip resistance and a black color, emery would be a way to achieve it. Another way would be to use a black pigment with a broomed surface. Your customer should be warned, however, that black surfaces may not remain a deep black indefinitely because of the tendency for lime and alkalis in the concrete to come to the surface and lighten the color.

Aluminum straightedges

Q. *Can I use aluminum channel successfully to make 10-foot straightedges for finishing concrete flatwork, or must I stick with magnesium? I know about the trouble with low strengths from entrained hydrogen gas when aluminum pipe was used to pump concrete, and I don't want to use aluminum if it's going to weaken the surface.*

A. We don't think there is any problem in using aluminum for this purpose. The aluminum is not in contact with the concrete for as long as aluminum would be if used in a pipeline, and any gas that does form should escape immediately from the concrete surface. So-called paver's straightedges made of aluminum are being sold, and we have not heard of any problems caused by using them.

Highway straightedge

Q. *What's a highway straightedge? Is there anything to the claims that it makes floors less wavy? If so, does it increase the labor cost?*

A. A highway straightedge, also known as a paver's straightedge, is a long-handled, T-shaped tool in which the crossbar of the T is made of metal tubing. The tubing may be aluminum or magnesium with about a 2x4-inch rectangular cross section, preferably plugged at the ends. It may be 10 feet long or more. The tool is used much like a bull float, except that the leading bottom edge cuts or planes the surface, depositing the cut material in the low spots. Its great length improves the flatness of the surface, whereas the bull float is said to contribute to waviness. The straightedge does not seem to increase the labor time once the operators get acquainted with using it.

Brooming

Q. *Why don't we get grooves and ridges of reasonably uniform width and depth when we broom a slab for slip resistance?*

A. It's not too much of a trick. Keep a wide, shallow pan of water at the edge of the slab. Before each pass, dip the broom in the water to wet it and remove any mortar clinging to the bristles. Be sure to tap to remove excess water, then use the broom.

Joints

Long-distance joint spacings

Q. *We would like to know if there have been any published data or surveys made of buildings where particularly long distances between joints have been built and proved effective or otherwise.*

A. Control joints were spaced 44 feet on centers in the Military Personnel Records Center, St. Louis, Missouri, a 282- by 728-foot, 6-story flat slab building built in 1956. The building was built with 22- by 22-foot bays with control joints at every second bay. Concrete was placed in checkerboard fashion, allowing 48 hours between placement of adjacent sections. Three-foot-long Number 4 dowels near the tops of the slabs cross the joints 12 inches on centers and the slab reinforcement also crosses the joints.

The General Accounting Office Building built in 1951 in Washington, D.C., an 8-story flat slab building 389 feet by 638 feet, built with 25- by 25-foot bays, has control joints 50 feet on centers.

No expansion joints were used in the Los Angeles Union Terminal 7-story flat slab warehouse, 100 feet by 550 feet, or in its 4-story 100 by 440-foot building, neither of which showed any distress from lack of expansion joints when inspected in 1958 at ages of 40 to 50 years.

The minimizing of expansion joints in the design of the first building is discussed in the article by Earl B. Cohn and W. A. Wahl, "Military Personnel Records Center Building Without Expansion Joints," *Journal of the American Concrete Institute*, June 1958, page 1103.

Half-round keyways and standardization

Q. *We manufacture wood shapes for chamfers, moldings, rustications, and various special uses. On a single day recently we had 3 separate inquiries for half-round keyways including a total of four different diameters: 1 3/8, 1 5/16, 2 1/4 and 4 inches. It is not only costly to tool up to produce any given diameter but it requires 4 to 5 weeks for delivery. This is frequently an impossible delay for the contractor. If some agency could standardize on a limited number of diameters, the cost of the keyways could be greatly reduced. It seems obvious that there would be little or no difference in performance between a 1 3/8-inch and 1 5/16-inch half-round keyway, which are only 1/16 inch different in diameter. But surely incremental differences in diameter could be considerably greater than that. Why not a joint design that makes use of only five standard sizes—1-, 1 1/4-, 1 1/2-, 1 3/4- and 2-inch diameters? If the designer were to specify only the total cross section that goes into shear the contractor could have his choice whether to use, for example, one 2-inch half round or two 1-inch half rounds.*

A. The use of half-round keyways is, as far as we know, fairly new and not widespread. The details shown here are taken from "Concrete Streets: Typical Pavement Sections and Jointing Details," *Concrete Information* IS211.01P, published by the Portland Cement Association, 5420 Old Orchard Road, Skokie, Illinois 60077. As noted in the lower detail, taken from that publication, the half round has a diameter of 0.2*d*, where *d* is the pavement thickness. According to this requirement, every inch of difference in pavement thickness leads to a difference of $^1/_5$ inch (approximately $^3/_{16}$ inch) in the diameter of the half-round keyway.

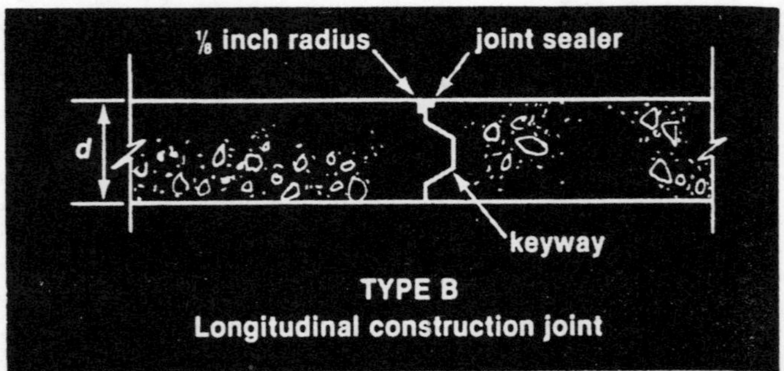

TYPE B
Longitudinal construction joint

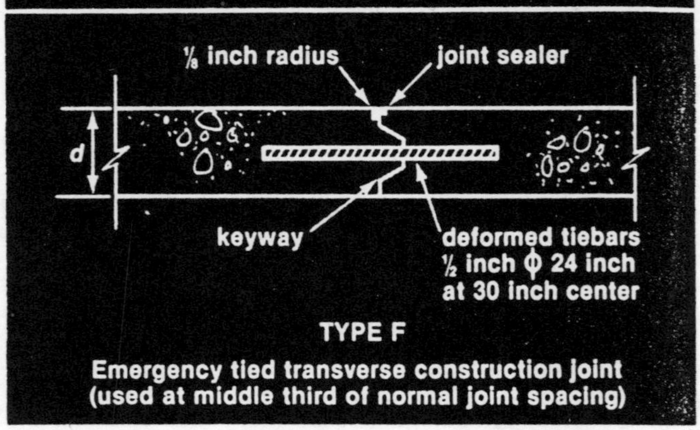

TYPE F
Emergency tied transverse construction joint (used at middle third of normal joint spacing)

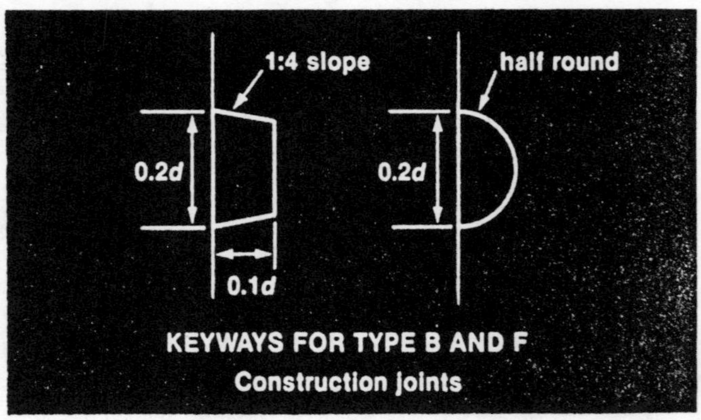

KEYWAYS FOR TYPE B AND F
Construction joints

Gordon Ray, of the Portland Cement Association, explained that the intention of this requirement, however, was to simplify the choice of keyway stock, not complicate it. It was anticipated that the contractor could choose the size of available half round that comes closest to 0.2*d*, which has been found to be the optimum size. A good rule, he says, would be to choose a stock size that falls within the range 0.2*d* to 0.25*d*. According to this rule, the diameters of half round to use for various pavement thicknesses are shown in the following table. The two center columns show the values for 0.2*d* and 0.25*d*.

Pavement thickness, inches	0.2d, inches	0.25d, inches	Diameter of half round to use, inches
10	2.0	2.5	2
9	1.8	2.25	2
8	1.6	2.0	1$^3/_4$
7	1.4	1.75	1$^1/_2$
6	1.2	1.5	1$^1/_4$
5	1.0	1.25	1 or 1$^1/_4$
4	0.8	1.0	1

Tool for jointing

Q. *How do I cut contraction joints across a wide slab of fresh concrete without using a bridge? We do outdoor flatwork and sometimes in hot weather the concrete cracks within a matter of hours after finishing. To avoid this trouble we want to form control joints immediately after screeding and bull floating. Walking groovers with long handles don't seem to be the answer because they dig in and are hard to manage at a distance.*

A. One contracting company, Northern Illinois Concrete and Paving Inc., of Sandwich, Illinois, devised its own tool, shown in the accompanying picture. It's a modified bull float that has a blade just off center on the bottom surface. This particular

blade can't be set at the center because its simple clamp won't accommodate the handle, but the company says it works fine in an off-center position.

The company bull floats the slab, then uses this tool to cut the fresh concrete, then bull floats again. They establish the joint line by snapping a string. They feel that the accuracy of locating the string line is more critical for driveways than for parking lots. When cutting across any joint previously cut at right angles, the operator has to be careful not to leave a ball of concrete in the older joint. The whole operation is said to require a good eye and a good feel for the use of the modified bull float.

The modern recommendation for depth of cut is $^1/_4$ the thickness of the slab. This tool, however, has a bit that is only about $^7/_8$ to 1 inch deep. It has nevertheless been successful in controlling cracking in slabs both 4 inches and 5 inches thick, perhaps in part because the cut is made so early. It is thought that with a deeper bit it would be hard to move coarse aggregate particles out of the joint when the bit is in a location some distance from the operator. The company says that for slabs 6 inches thick or more they rely chiefly on sawing.

At least one tool supplier offers a groover attachment for a Fresno trowel that looks much like the attachment in the picture.

Northern Illinois Concrete Paving used the tool shown here to joint a 4-inch-thick concrete overlay on an old 3-acre blacktop lot. Joints were located at 10-foot intervals in both directions. After one year there is reported to be only one random crack in the three acres.

When planning joint locations to control cracking, with this tool or any other, it's wise to divide the slab into areas that are as nearly square as possible.

Radius at top edge of saw cuts

Q. *Specifications for a food factory floor call for 15,000 linear feet of sawed joints $^1/_8$ inch wide and $1^1/_2$ inch deep and another 5000 linear feet of joints $^3/_{16}$ inch wide and 1 inch deep. They also require a radius at the top of the saw cuts. How can this be done?*

A. The radius requirement is highly unusual and unless there is some overriding good reason, we would recommend against it. If there is to be hard-wheel traffic on the floor the radius could contribute to spalling. If the joint is to be sealed, the radius would create an undesirable angle at the upper edge of the bonded surface, leading to sealant failure.

We do not know how a true radius could be readily produced by sawing. If absolutely necessary, the edge could be beveled by using V-shaped grinding tools, but it might not be possible to keep them from wandering without great expense.

Elongated slabs are likely to crack

Q. *What's the maximum recommended length-to-width ratio for sections between joints in a slab on grade? The drawings for a job I'm building show some bays that are twice as long as they are wide. No wire mesh is called for in the 6-inch-thick slab.*

A. The rule of thumb is that the ratio of panel length to width shouldn't exceed 1.5. This value is recommended in "Guide for Concrete Floor and Slab Construction," ACI 302.1R-80, in Section 2.3.2 on contraction joints. Panels with excessive length-to-width ratios are likely to crack at the mid-panel point or at some other location between joints.

The recommended joint spacing in feet is usually two to three times the slab thickness in inches. Thus, for a 6-inch-thick slab the maximum recommended joint spacing is 18 feet.

Don't push the limits when specifying joint spacing

In the Problem Clinic for July 1988, page 695, we said panel length-to-width ratio for slabs on grade shouldn't exceed 1.5, as recommended by ACI Committee 302. We also repeated a recommendation for joint spacing: the spacing in feet is usually two to three times the slab thickness in inches.

A consulting engineer called to warn about the dangers of designing joint spacings at the extremes of these limits. He contends that mid-panel cracking is likely in a 6-inch-thick slab with joints at 18 feet or in a panel with a 1.5 length-to-width ratio. In hot, dry weather or when poor construction practices are used, even 12-foot joint spacings for a 6-inch-thick slab may be excessive.

If panels do crack, the design may be criticized as unconservative in a claims action. The engineer had such an experience involving slab cracking. The contractor claimed the engineer was responsible for slab cracking because his design joint spacing "pushed the limits" of accepted practice. Our caller advised that having used excessive joint spacings to reduce joint sawing costs is a weak defense when the slab cracks. He said to be especially careful when designing thick pavements. The Federal Aviation Administration (FAA) recommends 25-foot maximum joint spacing regardless of slab thickness.

Joint filler problems in an industrial floor

Q. *We placed a concrete floor in an unheated 300,000-square-foot warehouse. Control joints were sawed on 20-foot centers each way. We filled the joints with epoxy but there has been quite a bit of joint movement and now the joint filler is pulling away from the concrete in spots. We've decided to remove the epoxy and fill the joints with something more pliable. Will the joint edges spall if we do this?*

A. Depending on the type of traffic over the floor you could have problems with joint spalling. Joint fillers should provide support at the edges, but a soft, elastomeric material won't give much support. Hard-wheeled traffic is likely to tear up concrete near the joint.

You don't say what kind of epoxy was used to fill the joints initially. It should be a semi-rigid material, sometimes called a flexibilized epoxy. There's a possibility that the gaps in the joint can be refilled without taking out the epoxy that's there now. This would help to accommodate the additional movement caused by shrinkage. Check with the epoxy manufacturers or a joint maintenance company to find out if this is feasible.

Low-flexibility sealants for floors

Q. *Many years ago I started filling industrial floor joints with lead strips, which seemed better than anything else available. In the past several years I've used mostly epoxies, because they bond to the sides of the joints. But they aren't the best answer, either, because they frequently cause a crack in the concrete parallel to the joint. What's the latest material, if any, that's better? Don't tell me "elastomers"; they're too soft.*

A. Some experts have recently been advocating semirigid epoxies for sealing joints in industrial floors. Generally an epoxy designed for the purpose, with a Shore A hardness of about 80, is said to be about right. These have a reasonable degree of hardness, yet also a small amount of extensibility. ACI Committee 302, in its "Guide for Concrete Floor and Slab Construction," recommends that such joint materials be 100 percent solids epoxy resins with a minimum Shore D hardness of 50 (ASTM D 676) and elongation of 6 percent (ASTM D 2240).

Shape factor isn't everything

Q. *Until reading your article "Pavement Joints That Last 5 Years?" (January 1983, page 33), I was convinced by lectors at concrete repair seminars and by manufacturers' literature that the proper shape factor for sealed joints was 2:1 (ratio of width to depth). This was to reduce volumetric change of the sealant for a given joint expansion or contraction. Your article (and detail shown on page 33) show the exact opposite as desirable. Who is correct?*

A. Frank Gaus, the author of the article, reports that if you are talking about the special sealants used for structural work, then a shape factor of 2:1 is desirable, as you say. However, Gaus says that in pavement work the inverse ratio of 1:2 works better. One reason is that shape factor isn't everything. In structural work it is possible to seal joints meticulously, and the proper width-to-depth ratio can readily be maintained. In pavement work the sealing cannot readily be done so carefully, and a sealant intended to be 1/2 inch wide and 1/4 inch deep may actually be so thin in some locations that it barely hangs onto the concrete from the moment the sealant is placed. Thus a shape factor of 2:1 provides far too little margin of safety. *It is very difficult in pavement work to get the full depth of sealant that is specified.*

Another problem, according to Gaus, is that in pavement work about 3 joints out of 10 do not open and close at all, leaving 7 joints to accommodate the movement expected of 10, and even these 7 joints will not divide their total movement equally among themselves. Within any group of joints it often happens that most of the movement is being taken by a single joint, as shown in a paper "Magnitude of Horizontal Movement in Jointed Concrete Pavments," by I. Minkarah, J. P. Cook and J. F. McDonough, published in *Transportation Research Record 821*. They also found that it is not always the same joint that takes the large movement.

The deeper reservoirs that Gaus recommends for pavement joints provide not only more material but a larger bonded area. He recommends the following ratios, depending on the width of the joint:

Width, inches	Depth, inches	Ratio, width:depth
1/2	1	1:2
3/4	3/4	1:1
1 1/2	3/4	2:1

More on shape factor of joint sealants

Sherwood Spells, Technical Representative for Dow Corning in Midland, Michigan felt the information from Frank Gaus wasn't complete in the answer to "Shape factor isn't everything," page 357 of the April issue.

You gave information from Frank Gaus in answer to a question which was essentially: "What is the correct joint sealant design, a 2:1 or a 1:2 ratio of width to depth?" The reply stated that a 1:2 ratio works better for pavement sealing than a 2:1 ratio. Several reasons for this recommendation were cited. However, this reply addressed only half of the story. It is the other half that I wish to comment upon.

A ratio of 1:2 will produce a deep sealant bead. As shown by Egons Tons in his classic article "Geometry of Simple Joint Seals Under Strain" (*Highway Research Board Bulletin 229,* 1959), deep sealant beads place more strain upon the sealant. Thus, shallower sealant beads are preferred over deeper ones. Although Ton's work was centered around materials available in the late fifties, the concepts are still valid today with newer sealing materials.

There are several manufacturers of generic sealants (silicones, polyurethanes, polyvinyl chlorides, and others). Each generic sealant has different handling and performance characteristics and installation requirements. These unique characteristics, such as weatherability and adhesion, along with extensive laboratory and field testing help determine the proper joint design for any given generic sealant. Some sealants have proven field performance when shallow sealant bead is used, whereas others require a deeper bead. Thus an acceptable joint design for one sealant may be unacceptable for another sealant. The manufacturers recommendations should be followed.

As Frank Gaus states, installing sealants to a recommended joint design in structural work is not difficult. Similarly, it is not difficult to install highway sealants according to joint design. It's a matter of tools and techniques. All field molded sealants require a bond breaker to prevent three-sided adhesion as well as to control the sealant depth. The bond breaker is installed first to the proper depth, followed by the sealant. Thus the desired sealant shape is obtained. Obviously, there are a few examples where the sealant bead is too shallow. However, these are the exceptions rather that the rule, as Mr. Gaus implies.

Expansion joints

Q. *Could you please let us know where we might obtain up-to-date information on expansion joints in concrete buildings?*

A. There is an excellent discussion in *Handbook of Concrete Engineering* by Mark Fintel (published by Van Nostrand Reinhold), pages 101 to 103. It discusses their purpose, places where they are used, spacings between joints as related to expected temperature range, the need to include them in the preliminary planning stage of design, how they function, factors that affect design, and joint details.

Joints in tennis courts

Q. *A project consisting of three tennis courts and supplementary aprons is scheduled for bidding. The job is now set up as a blacktop paving project but the engineer will accept an alternate proposal for a concrete court system provided he can be assured of how the construction joints will be built so as not to have a ridge occur between one pour and another.*

Are there any standard methods, recommendations, or drawings as to how the joints should be constructed, and where they should be located?

A. It is the recommendation of the Portland Cement Association that no joints be used within the court itself, but that enough steel be used to keep hairline cracks tightly closed. For this purpose they recommend using Number 5 reinforcing bars at 12 inches on centers in each direction for a 5-inch-thick slab, or at 15 inches on centers for a 4-inch-thick slab. These should be located at mid-depth of the slab and be supported every 4 inches by bolsters or every 3 inches by concrete bricks. If necessary to construct the court in two parts, an expansion joint can be used under the net. Expansion joints should also be used between adjacent courts.

Information is provided in *Reinforced Tennis Courts*, Publication IS199T, available for purchase from the Portland Cement Association, 5420 Old Orchard Road, Skokie, Illinois 60077.

Curing

Restarting the curing process

Q. *If concrete dries out before the specified curing period is over, can water be applied to restart the curing process?*

A. When moist curing is interrupted, hydration of the cement continues for a short time and then stops. If water is applied later, hydration and strength development will resume as soon as the concrete reaches a high enough moisture content, provided that the temperature is high enough (above about 50° F). Because it may be hard to get enough water back into the concrete, it's best to cure the concrete continuously from the time it's placed until it reaches the desired strength.

Cost to fog spray concrete flatwork

Q. *We're bidding a parking deck job that will be built with silica-fume concrete. The weather is likely to be hot and windy so we're worried about plastic shrinkage cracks. To prevent cracking we might use a fog spray until finishing is completed and a curing compound can be applied. For bidding purposes, however, we don't know how to estimate the cost of fog spraying. Can you tell us what the costs might be?*

A. We didn't have much luck getting an answer to this question. One floor contractor says the cost might be 5 to 10 cents per square foot, but that this was just a guess. Fogging nozzles cost about $5.00 each. They deliver a fine spray (1/2 gallon per minute) and require a minimum water pressure of about 30 psi. At a distance of 6 feet above the surface the nozzles produce a spray about 6 feet in diameter at the surface. For wide pours it might be necessary to build a movable pipe frame and mount the nozzles on it. For smaller pours, workers on each side of the slab could use garden hoses with fogging nozzles to keep the surface from drying.

Fog spray costs

In response to our request for fog spray cost information (Problem Clinic, July 1987, page 634), R. A. Prisby and Mark Luther of Elkem Chemicals provided the following information:

The cost of fogging varies greatly depending on equipment on hand, placement size, project size, weather conditions (wind, temperature, humidity), and number of placements exposed to the weather (top decks). When fogging is needed, base your costs on using one person who has no other responsibilities than to operate the fogger(s). Fogging is a full-time job on days when it's necessary.

On large projects one-time costs usually represent a small portion of the total cost of fogging. There also may be proportionately few placements that require fogging. One estimate for a medium-sized project (a 270,000-square-foot parking garage requiring extensive fogging for 30 percent of the placements and assuming no equipment on hand initially) puts the fogging cost at $0.011 per square foot. For larger projects the cost is lower; for smaller jobs the cost may be considerably higher.

Other options are available, however. One is the use of evaporation retarding materials as supplied by the major admixture firms. These chemicals have successfully prevented plastic shrinkage cracking with silica-fume concrete. Perhaps the best option, though, is to place the concrete at night or delay placement until weather is better.

We have seen mixed results with fog nozzles attached to garden hoses. The water pressure generally is not high enough to produce a fine mist. However, we have seen success with nozzles that incorporate compressed air to blow the water out. An effective nozzle assembly is shown in Figure 1.

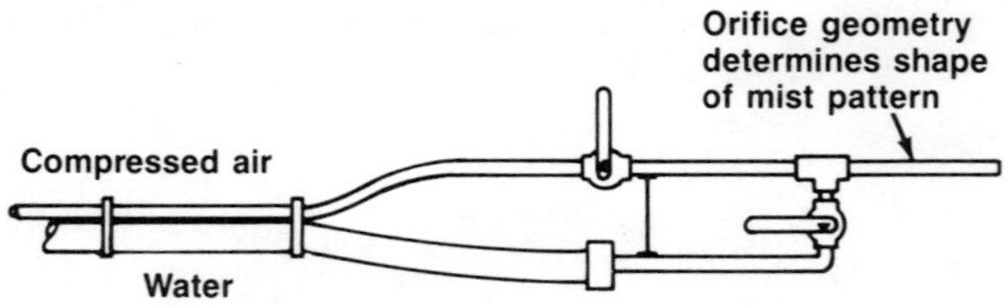

Figure 1. *This fogging apparatus uses compressed air and water to create a mist fine enough to help prevent plastic shrinkage cracking in silica-fume concrete.*

Percent solids in curing compounds

Q. *What's the significance of percent solids in liquid membrane curing compounds? The cure-and-seal products seem to have a higher solids content. If two products meet the ASTM requirements for curing compounds, is the one with a higher percent solids more desirable?*

A. We asked Bill Phelan of Euclid Chemical Company to answer this one. He says high-solids curing compounds (about 30% solids) won't permit as much moisture loss. Curing compounds meet ASTM C 309 requirements if, at a coverage rate of 200 square feet per gallon, they hold moisture loss to 0.55 kilograms/square meter of surface in 72 hours. A high-solids curing compound at a coverage rate of 300 square feet per gallon typically allows a maximum moisture loss of only 0.30 kilograms/square meter. Curing is better because of the lower moisture loss.

Another advantage, according to Phelan, is that high-solids compounds are less likely to be applied too thin. It's easy to tell when the coverage is correct because the compound leaves a detectable gloss on the concrete surface. One caution: cure-and-seal curing compounds are excellent bondbreakers; don't use them if tile or other coverings are to be bonded to the floor.

Curing compound for calcium aluminate cement

Q. *Can a curing compound that meets the ASTM C 309 specification be used with a calcium aluminate cement? This type of cement is not listed in the Bureau of Reclamation's* Concrete Manual, *Eighth Edition.*

A. As long as the curing compound is satisfactory for portland cement concrete there is no reason to believe that it would be unsatisfactory for curing concrete made with calcium aluminate cement. However, it is important to note that proper use of curing compounds cannot be overemphasized when used on this type of concrete. Hydration of calcium aluminate cement takes place almost entirely in the first 24 hours. The concrete must be protected from loss of water during this period.

Cure shotcrete thoroughly

Q. *Some unusual job scheduling is going to require us to apply shotcrete in three layers applied several days apart. We know we'd better not cure the first two layers with curing compound since it would probably interfere with bond. Is covering with polyethylene satisfactory if we make all edges tight?*

A. Shotcrete usually has a quite low water-cement ratio, and it can benefit by having curing water added. This will ensure that water will find its way into all the pores that might otherwise become partially desiccated by the hydration of the cement. Some means should be provided to keep the shotcrete moist between times—for example by fog spray or by water running down over it from soaker hoses. The final coat should be kept continuously wet for 7 days.

Paving

5

Concrete

Subgrade and subbase

Q. *I frequently see the terms subbase and subgrade, and they confuse me. Are they the same or different? If different, which one goes on top?*

A. In concrete paving terminology, the subgrade is the soil the pavement is built on. Usually it is treated by compacting it sufficiently to prevent the pavement from settling. Subbase is composed of the layer(s) of material between the subgrade and the concrete pavement. A subbase may not always be required. The subbase may be econocrete, cement-treated subbase, soil-cement, asphalt-treated subbase, crushed stone or other suitable material.

Acceptable thickness in parking lot

Q. *What tolerances have been established for the thickness of parking lots? We have a 5-inch pavement that averages about 0.35 inch less than specified thickness (measured by ASTM C 174). In one place the deficiency in thickness is more than 1.5 inch.*

A. We don't know of any standard tolerances for parking lots. Various state highway departments have thickness tolerances for highway pavement slabs. Some highway specifications indicate a range of deficiencies in thickness that are considered acceptable (within tolerance limits); in another, higher range deficiencies affect the amount of pavement; in a third range, higher deficiencies may be sufficient to cause rejection of the pavement. Most states and Canadian provinces have established their own standards. Some of the more stringent specifications allow only 0.01 foot deficiency in thickness for an acceptable pavement, but 0.2 inch is more typical and some values are as high as 0.5 inch. Some states impose a penalty if deficiencies are above these values. The amount of penalty is sometimes written into the job specifications. There are other higher values (0.5 inch in some states but up to 1 inch in others), above which no payment will be made. In such cases the pavement may have to be torn out and replaced. These and many other details on pavement specifications may be found in "Charted Summary of Concrete Highway Pavement Practices in the United States," IS201P, available from Portland Cement Association, 5420 Old Orchard Road, Skokie, Illinois 60077.

Saw pavements promptly

Q. *What caused a crack to form early in our recent parking lot construction? We used air-entrained concrete estimated to attain 5500 psi compressive strength and placed it at a slump of 5 inches and finished it with a vibrating screed. The air temperature was 80° F or slightly higher. When the slab, which was 5 inches thick, was between 21 and 23 hours old we sawed it into 12-foot-square panels by cutting $1^1/2$ inches deep. By then the slab had already cracked in one place on a line about 1 to 2 feet from the intended location of one of the joints. This crack extended across a 24-foot width of pavement. As a result we decided to omit the joint that we had intended to cut at that location.*

A. The general rule is that slabs should be sawed as soon as sawing can be accomplished without causing the joints to ravel. The ACI 302 Guide for Concrete Floor and Slab Construction says that at normal temperatures this can usually be done by sawing within 4 to 12 hours after placing. Since you placed the concrete in warm weather, it seems obvious that you were sawing at a dangerously late time.

Assuming you do not plan to repair the crack (for example by epoxy injection), it was a reasonable decision not to saw a joint adjacent to the crack. That would have produced additional irregularly shaped small narrow panels. The existing panels adjacent to the crack should function much like the other panels since they are of comparable sizes. A disadvantage is that a crack is more likely than a joint to undergo progressive deterioration. Another obvious disadvantage is that the crack does not look as neat as the joints and would be more difficult to seal.

Transfer load between new slab and old

Q. *We must replace some large areas of highway pavements that contain large, irregular cracks that open and close, some of which intersect. If we remove the old concrete how can we be sure the new concrete won't settle independently or tilt to form faulted joints?*

A. Saw out large areas to eliminate cracks. Use a diamond core drill to make horizontal holes in the adjacent concrete and insert load transfer dowels across all the new construction joints. These should be cemented in place with a polymer concrete or other proven patching compound. Place concrete for the new slab, consolidate it by vibration, finish it to a texture like that of the surrounding concrete, and cure.

Admixtures from recycled aggregates

Q. *We are considering purchasing concrete from a plant that uses recycled aggregate. This job will continue for some time, so we are concerned about whether the plant will be able to deliver the same quality of concrete throughout the job. What worries us is that the recycled aggregate may contain all kinds of things—retarders, accelerators, water reducers, air-entraining agents—one kind on one day and another kind or two on another. What is this going to do to our concrete? Won't it sometimes be accelerated, sometimes retarded or sometimes air-entrained, depending on where the recycled aggregate comes from?*

A. The effects of the admixtures in the recycled aggregates should be almost negligible. As a matter of fact, two Danish researchers, Torben Hansen and Soren Hedegard, have reported an interesting investigation of this subject in the January-February 1984 *ACI Journal* (pages 21 and 26). They made some concretes that were identical except that one contained calcium chloride accelerator, one a retarder, one an air-entraining agent, one a superplasticizer, and one of them no admixture at all. After these concretes had been cured for 39 days, portions of them were broken up and made into aggregate of about the same gradation as the original. The 5 aggregates from the 5 concretes were then used to make concretes of approximately the same mix proportions as the original nonadmixtured concrete. It turned out that the investigators could see little difference among these recycled concretes regarding their setting times or strengths. Nor did the recycled aggregate from the air-entrained concrete have any significant effect on the air content of the new concrete. The aggregate containing calcium chloride did raise the chloride content of the new concrete, but not enough to account for all the original chloride. Further details are given in the article cited.

Recycling D-cracked pavements

Q. *Now that highway departments are beginning to recycle concrete pavements by crushing them for use as aggregate in new concrete, we wonder if it is possible to recycle pavements that have been subject to D-cracking. Would they just begin to break up again in the same old way?*

A. D-cracked pavements have been recycled but it may be too early to know how successful the experience has been. If there is any doubt about future susceptiblity to D-cracking it would be well to test the materials concerned. The test method is given in a report by Paul Klieger, Gervaise Monfore, David Stark and Wilmer Teske, "D-Cracking of Concrete Pavements in Ohio," Report Number OHIO-DOT-11-74, National Technical Information Service (NTIS), U. S. Department of Commerce, Springfield, Virginia. (Soft cover, 201 pages)

The first major concrete recycling project in this country to reclaim pavement susceptible to D-cracking was a 16-mile project on U. S. 59 between Worthington and Fulda, Minnesota.

Grooving for skid resistance

Q. *The concrete pavements in the central part of our village are serving remarkably well in spite of their 1940s vintage. We don't have much cracking or faulting, but they are worn smooth in the most heavily trafficked areas. We are thinking of safety grooving. Does grooving have to go across the pavement to give good braking and resistance to skidding, or can it run lengthwise, as on expressways? How far apart should the grooves be spaced?*

A. Longitudinal grooving is effective in improving both braking and cornering friction coefficients. It is reported to make a wet pavement perform almost as well as a dry pavement. You can get a variety of recommendations on groove spacings, depending on whom you consult, but there is good evidence that spacings ought to be $1^{1}/_{2}$ inch or less.

Pavement grooving and texturing

Peter Silveri of the International Grooving and Grinding Association had advice for those with questions on grooving:

In the Problem Clinic on page 1008 of your November 1984 issue I noticed a question in regard to grooving for skid resistance. If readers of your publication want more detailed information on pavement grooving and texturing, we would be happy to assist them. They should write to us at: International Grooving and Grinding Association, P.O. Box 1750, Skyland, North Carolina 28776; telephone 704-684-1989.

Longitudinal or transverse grooving

Q. *We built a pavement that was supposed to be textured by transverse tining for skid resistance and good braking. However, we got caught by rain, which washed away the existing texture and prevented further tining. We are now required to groove it transversely. This requires closing two lanes, is a danger to the public, and is very slow. We would like to groove the pavement longitudinally but the state allows only transverse grooving. Do you have any information on the relative effectiveness of longitudinal and transverse grooving on skid resistance and braking?*

A. Walter B. Horne has written a technical report for the International Grooving and Grinding Association called *Safety Grooving, Hydroplaning and Friction*. In that report he says "Many highway authorities have long been puzzled by the seeming paradox that while longitudinal safety grooving does greatly reduce wet weather skidding accidents, friction coefficient measurements made before and after grooving often show little change. This result has led to the belief by many that grooving reduces hydroplaning but does not improve the coefficient of friction." However, he then points out that most highway departments measure friction coefficient by an ASTM standard method that requires the skid trailer to lock the test tire while traveling over the pavement at 40 miles per hour. He cites data from tests by the Pennsylvania Transportation Institute that show that, if the wheel is braked without locking, the friction coefficient is higher. Thus the braking effectiveness should be higher than most measurements made by the standard ASTM method indicate. Some of the data are also shown in Figure 4 of our article, "Skid Resistance Theory Related to Grooving Practice," published in the August 1983 issue, page 619.

Concrete saw blade life

Q. *We're saw cutting a road that consists of 6 inches of concrete overlaid with 1 inch of asphaltic concrete. In most places there's only $^{1}/_{2}$ inch of asphalt left. The cut is 3 inches deep and our saw blades are lasting for only 500 lineal feet of cutting. Isn't that low? We're using a 14-inch-diameter asphalt-concrete blade with a 26-hp saw. Cooling water is gravity fed from a flatbed truck through a $^{3}/_{4}$-inch-diameter hose.*

A. We asked several concrete saw manufacturers and concrete sawing contractors for their suggestions. They said blade life depends a lot on the concrete you're cutting. Hard aggregates such as river gravels wear out blades faster than soft aggregates such as crushed limestone. Thus it's hard to say whether or not the 500 lineal feet is exceptionally low. Also, there are as many as five different qualities of blades, with top-of-the-line blades lasting much longer than lower-priced blades. Buy the best blade available. It might cost 30% more but can give you as much as twice the cutting life.

It's better to cool the saw blade with pumped water. You need $2^{1}/_{2}$ to 5 gallons per minute, and it's tough to get that much water with gravity feed through a $^{3}/_{4}$-inch-diameter hose. If smoke or steam is coming from the saw cut you're not getting enough water to the blade and blade life will be reduced. Also, try using an 18-inch-diameter blade if the saw will take the bigger blade.

Roller-compacted street pavements

Q. *We've read about roller-compacted concrete (RCC) being used for streets and roads. It's reported to cost less than asphalt pavements. What's the smallest size job that can be done economically?*

A. Ron Pelz of Pelz Construction Company in Alliance, Nebraska, says that he can bid competitively on some jobs requiring as few as 5,000 cubic yards of RCC. However, to get maximum savings for an RCC job, a volume of 20,000 cubic yards or more is best. You'd need about 5,000 cubic yards of RCC for a mile of pavement 40 feet wide and 8 inches thick.

1/2 plus 1/2 doesn't equal 1

Q. *We're going to be paving a parking lot that is 90 by 220 feet. The plans call for a 1/2 percent slope in both the east-west and north-south directions. I told the owner's representative that to avoid birdbaths the minimum slope should be 1 percent (1/8 inch in 1 foot). The representative said a 1/2 percent slope in each direction is equivalent to a 1 percent slope. How do I deal with this kind of logic?*

A. You're right about the 1 percent minimum slope needed to prevent ponding or birdbaths. The Portland Cement Association (PCA) publication *Building Quality Concrete Parking Areas* recommends sloping the pavement 1 percent to drainage. The American Concrete Institute's (ACI) *Guide for Concrete Floor and Slab Construction* recommends a slope of about 2 percent (1 inch in 4 feet) to provide positive drainage.

Here's the logic we'd use to show that 1/2 plus 1/2 doesn't equal 1. Take a one-foot-square section of the lot sloped 1/2 percent in each direction. The greatest fall from the high corner to the low corner will be 0.01 foot (0.005 foot in the east-west direction plus 0.005 foot in the north-south direction). Water will drain on a 45° angle across the section from the high corner to the low corner. The diagonal has a length of 1.414 feet so the slope is (0.01/1.414) x 100 = 0.71 percent.

We'd recommend using a minimum 1 percent slope in both directions even though the resulting slope in the direction of water movement will then be slightly greater than 1 percent.

Colored traffic lanes

Q. *While planning the offramps for an expressway the question has come up whether these can be colored to distinguish them from the regular traffic lanes. If so, can you suggest how much color to use?*

A. This is a technique that has been used successfully in a number of places. Often red pigment is used in the mix. When a three-mile stretch of First Avenue in New York City was paved in 1983, one lane was colored black to help delineate it for the exclusive use of buses. In that instance 28 pounds of carbon black pigment was used per cubic yard of concrete containing 588 pounds of cement and designed for 4000-psi strength. The concrete was supplied by a ready mixed concrete producer. We suggest you talk with such producers or with manufacturers of pigments to get their recommendations.

Parking lot white top

Q. *Is it cheaper to pave with concrete over existing blacktop than to do a complete new concrete parking lot? The existing blacktop is somewhat wavy and not uniform in elevation.*

A. Easy Street Construction Company, of Naperville, Illinois, has done both kinds of paving and prefers to work on new ground if it is hard clay. They say that if the job is on new ground and can be prepared with a trimmer, calculating the concrete costs is straightforward. But on an uneven blacktop base it is uncertain how much concrete will be needed. Hence the contractor needs to give both a square-foot and a cubic-foot price. Even if an engineer has drawn up the whole design the contractor cannot know exactly how much concrete is needed. On all rebuild work the contractor needs a protection clause for quantity of concrete. On new work, on the other hand, the contractor can inspect the subgrade and figure concrete quantities based on the company's known tolerances for trimming the subgrade and placing the concrete. Easy Street says that when they estimate 6 inches thickness on a trimmed subgrade they get 6 inches.

Standards for concrete pavement for parking lots

Q. *This is my second year as a contractor doing residential and some commercial work, primarily walks, driveways and parking areas. In looking for recommendations on what the standards ought to be I've run into a lot of different opinions from people I thought would know the answers. I'd like to know:*

1. *Can I really get by with 4 inches of concrete for a parking area?*
2. *Should I use wire mesh?*
3. *Should there be a layer of gravel under the pavement? I'd like to know I'm doing the job right, but I don't want to waste money on things that aren't necessary.*

A. The different opinions may reflect some changes in standards that are taking place. For many years asphalt has dominated the parking lot market, for no better reasons than its being cheap and easy to apply. Concrete was almost always overdesigned for the purpose and therefore more expensive. In recent years the concrete industry has made a concerted effort to capture a larger share of that market. This has resulted in the application of more realistic minimum standards that are involved in all three of your questions.

1. A 4-inch thickness is more than is needed for the typical parking lot traffic of passenger cars and light pickup trucks. However, it is a practical minimum for construction. If the traffic pattern will bring heavier trucks near the free edge of the slab (at the slab perimeter or at an expansion joint) thickening the slab edge gradually from 4 inches to 6 inches over a distance of 36 inches from the edge would be a good precaution against overloading.

2. You will not usually require wire reinforcing mesh. It doesn't increase the load carrying capacity of the slab; its primary purpose is to prevent the opening of shrinkage cracks that occur between joints. In a nonreinforced 4-inch slab, control joints should be spaced at a maximum of 8 to 10 feet and cut to a depth of 1 inch. If that is done, the shrinkage is provided for, and there is no need for wire mesh. If it is desired to space joints farther apart than 10 feet, mesh would be helpful.

3. A granular subbase is used to prevent pumping at joints in concrete highways subjected to heavy loads moving at high speeds. Since these conditions do not exist for driveways or parking lots, a granular subbase is not required. The recommendation is merely that the subgrade be uniformly compacted. Apart from concrete pavement design, however, there are conditions under which some granular material may be helpful in construction. Heavy clay soils can be difficult to grade accurately. It may be economical to bring in a couple of inches of granular material to allow more accurate fine grading, especially when working with a minimum pavement thickness. Some granular material may also provide a better working platform in case of a wet subgrade.

Don't forget some other important standards about which you didn't ask: 3500 to 4000 psi concrete, air entrainment in severe climates and a maximum slump of about 4 to 5 inches.

Pattern cracks in parking lot

Q. *We built a concrete parking lot in Chicago 6 years ago. It's in excellent condition except for one area that appears to be progressively deteriorating. There are a couple of shrinkage cracks in this area and these are surrounded by a system of smaller cracks that seem to be filled up with a gray substance. The cracked area gets a little larger each year. The concrete was air entrained and we placed it with a low slump and vibratory screed. There's no scaling at all. What could have caused the cracks?*

A. It might be D-cracking. D-cracks are a series of fine cracks that form at close intervals (see photo next page). In slabs on grade they are usually parallel to edges, joints, or other cracks. The cause is usually porous coarse aggregate particles that become saturated and then freeze during cold weather. Initially the particles crack internally, but as time goes on cracks extend into the surrounding mortar and to other aggregate particles. Sometimes the cracking spreads several feet beyond the joint or crack, and damage is severe enough to require replacing the affected concrete. The affected pavement section may still last 15 to 20 years before replacement is needed. Because it appears in only one section of the lot, aggregate in the other concrete may be of better quality and not susceptible to damage. We'd advise watching the section for a few more years to see if the cracked area continues to get larger. As long as the concrete still carries traffic loads there's no need to replace it unless appearance is objectionable.

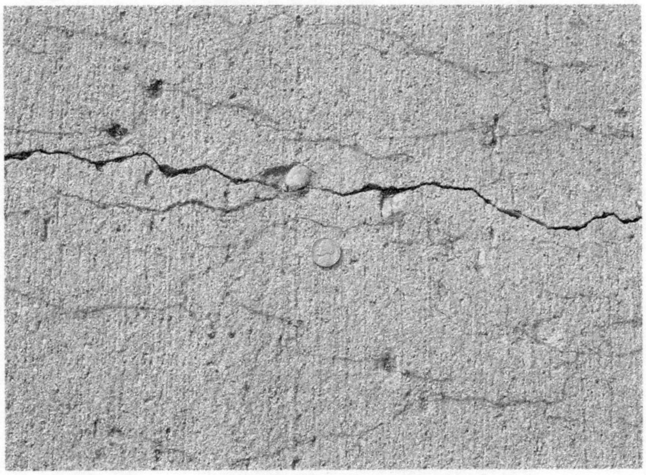

D-cracks frequently occur parallel to pavement edges, joints, or other cracks.

Pavement buckling

Q. *Why do we occasionally run into concrete pavement buckling on hot summer days? So far as we know there is no trouble in our area with expansion from reactive aggregates.*

A. Theoretically this shouldn't happen because, after normal drying shrinkage, concrete seldom becomes larger than its size when cast. Buckling does happen in practice, however, and the usual cause is that foreign material has been worked into the joints so that they no longer function properly. The foreign material could be pebbles that have penetrated the joint sealant or dirt and debris that have intruded after joint sealant has been torn out. The more wide open the joints when they become filled with foreign material, the greater the stress when they try to close in hot weather. Ultimately this causes some joints to buckle or "blow up" in the hottest weather. The problem seems to be worse on pavements that have longer distances between joints. Prevention is a matter of keeping joints well sealed and clean.

Troubleshooting

6

Concrete

Too much air

Q. *A new slab on grade in a parking garage was found to have 12 percent entrained air, rather than the 6 percent that had been specified. We don't question the accuracy of the air test but we want to know how this air content will affect the strength, durability and integrity of the job.*

A. As a general rule of thumb, the compressive strength of concrete will be lowered by 3 to 5 percent for each 1 percent of entrained air. Air-entrained mixes are proportioned with this in mind. A mix with an unexpected excess of 6 percent entrained air could have a compressive strength 18 to 30 percent below design strength. Abrasion resistance and general durability will also be markedly lower than expected.

Short of replacing the slab, the only corrective measure would be to cast a topping over it. The type of topping and the thickness would be determined by performance requirements and how much increase in floor elevation would be acceptable.

Honeycomb tolerances for sandblasted surfaces

Q. *We have recently had stringent inspections of architectural concrete with a sandblasted finish. Many of the defects relate to honeycomb revealed by the sandblasting. Please send any literature you have about tolerances for percent defective area allowable.*

A. We do not know of any established tolerances. In architectural concrete work it is usually best to first do some panels before actual construction begins and then to finish the surface by the technique proposed for the job. Sometimes more than one technique may be used. The amount of honeycomb showing will be influenced by the placing and vibrating techniques as well as the amount of sandblasting done.

By conferring with the architect about the results it is possible to agree on what visual effect is expected by the architect and possible to execute. It is then a good idea to do a mockup to be sure the effect can be accomplished on the same scale as the final job. This procedure is usually more satisfactory than adopting someone else's tolerances, which possibly neither the architect nor the contractor will be happy with.

Slow strength buildup can be good

Q. *We placed some concrete that was supposed to reach 2000 psi compressive strength at 28 days and 1000 psi at 7 days. It didn't meet either requirement but sometime between 40 and 50 days the strength surpassed 2000 psi. Since that time we have pinpointed the problem and corrected it. However, now the question arises, is concrete that reaches 2000 psi at about 50 days as good as concrete that reaches 2000 psi at 28 days?*

A. There are really two possible questions here:

1. If two concretes reach a strength of 2000 psi at 28 days and 50 days respectively, are they equally good *at the times they reach that strength?* The answer is a qualified *yes.* They both have the same strength, though there may be minor differences in absorption and permeability.

2. If two concretes reach a strength of 2000 psi at 28 days and 50 days respectively, will they be equally good *at some later age?* The answer depends on the conditions.

(a) If the reason for slow strength gain in one concrete is that it was made of Type II or Type IV cement, or that it contained too much retarder, or was cured at a low temperature, it may actually end up being a better concrete than the other. It may develop more ultimate strength and better properties in general. The key factor is that if the curing conditions are such that the cement has plenty of time to hydrate, the newly forming cement gel will deposit itself uniformly throughout the matrix. When cement must hydrate rapidly, the cement gel tends to deposit itself on and around the original cement grains. This eventually tends to choke off these grains from reacting with the surrounding water. It also causes the gel to distribute itself less evenly so that the whole matrix is weaker. Slow, steady curing, whether achieved by low (but above-freezing) temperature, by a retarder, or by a slow-hydrating type of cement, is likely to produce a higher ultimate strength, other things being equal. One place where the effects of temperature and rates of curing are discussed is in A. M. Neville's book *Properties of Concrete,* John Wiley and Sons, New York, 1973, pages 276 to 280.

(b) If, however, the reason for the slow gain in strength is that the mix was improperly proportioned, or contains deleterious impurities such as clay, or was not cured properly but was allowed to dry out, or was abused in some other way, the concrete will not gain its full strength potential. If so, the concrete *is not* as good as the concrete that gained strength at the rate intended.

Wood embedded in concrete

Q. *I am involved with the design of an earth-sheltered home. Some of the concrete walls are to contain strips of wood embedded in the surface for the purpose of attaching brick veneer. I am concerned that the wood may soak up water, expand and rupture the concrete. Do you have any information on this?*

A. A brief item, "Wood Swells," in our October 1978 issue, page 584, recommends against using wood embedded in concrete because wood swells whenever it soaks up water. Embedding wood in locations where joints or cracks would be visible, as in the walls of an earth-sheltered house, is undesirable.

We should note that wood strips *are* sometimes embedded in concrete to form control joints in sidewalks, driveways and patios. This is a different *function.* It allows movement of the concrete at the joint, and some swelling of the wood can be accommodated there without damaging the concrete. Wood used for such a purpose is usually a rot-resistant wood such as cypress, cedar or redwood, usually primed with a wood sealer.

Concrete contaminated by urea

Q. *We noticed the smell of ammonia in fresh concrete mixes that were made with cement from several cement hauling trucks. In checking back, it appears that when these cement trucks were not being used for hauling cement they had been used for hauling urea fertilizer. We suspect that the urea was not cleaned out of the trucks before cement hauling. We have made trial mixes and as far as we can tell they behave like mixes that have not been contaminated at all. Is it safe to use this cement in concrete? If not, are there some precautions we can take so that we may be able to use it?*

A. We do not know of any published information on the effect of urea mixed into concrete but the chemistry of urea is such that in small quantities it could not do any harm to the concrete. Inasmuch as there has not been any noticeable change in the setting time of your trial mixes (which could happen if the urea had formed significant amounts of carbonates) the urea does not seem to have had any chemical effect. The small amount of contamination is not likely to do any harm. This does not, however, justify the use of cement trucks without thorough cleaning if they are used for hauling some foreign material before hauling cement.

Leakage in earth-sheltered house

Q. *We have a 32- by 120-foot earth-sheltered house built of precast, prestressed slabs, with a 2-inch thick cast-in-place deck over the slabs, coated with two coats of waterproofer, insulated with 4-inch polystyrene slabs, covered with polyethylene with only one seam, then covered with 2½ feet of earth. Walls are reinforced concrete masonry. There is slight leakage on the backwalls, mainly where the precast slabs meet. Is it possible that the problem could be remedied without removing the earth?*

A. This may be a possibility depending on the nature of the leak. There is a good new and fairly simple method of repair by epoxy injection that does not require a pump. It may work if the masonry walls have been filled solid.

Temperature problems with shotcrete

Q. *Our company builds hundreds of wet-mix shotcrete pools a year in the Southwest, using a pea gravel mix that we buy from either of two ready mixed concrete suppliers. Recently we have had a number of failures in which the shotcrete has crumbled away. This happened when the weather was very hot and I believe the problem is due to high concrete temperatures and maybe to hot cement in the mixes. We have never had this trouble in previous years when we used ice as part of the mixing water. Could hot cement be the problem?*

A. If hot cement has an effect it is a minor one. Even if the cement being used were at a temperature of 185° F it would raise the temperature of the final mix only a few degrees F. (This is because the cement is only a fraction of the mix and because the cement has a much lower specific heat than water.) We suggest you go back to using ice as part of the mixing water.

False set

Q. *Explain to me again what you mean by false set. Is it the same as flash set?*

A. False set is what it says—not the real thing. If freshly mixed cement, mortar or concrete begins to get rigid, without giving off heat, and if it can be made plastic again by continued mixing without adding more water, you have false set. If you have trouble with false set, mix the concrete longer, but also talk to your concrete or cement supplier. Some people add a chemical admixture to keep false set from happening before the concrete is in place.

Flash set is something else; the concrete gives off heat, and the set is permanent. If you have *that*, your concrete supplier or cement company should be able to correct it.

Excessive retardation

Q. *Part of a 4½-inch-thick concrete deck that is to be post-tensioned is not hardening, though it is 19 hours old. Only one truckload is involved, and the troublesome area is surrounded by good concrete. All the concrete contains a retarder. Can we do something to make it set or must we take it out?*

A. It sounds as though you've had an overdose of retarder in that truckload. This can happen through a batching error or when an admixture dispenser malfunctions. Can you wait a few days? The concrete will most likely set up if given enough time and should ultimately gain its full strength. If the concrete never does set (which is unlikely) you may want to get a testing laboratory to find out what got into the mix to retard it so badly. If it does set in a few days it may be wise to core the slab at age 28 days to see if the proper strength has developed.

Unwanted surface retardation

Q. *I built a patio in my backyard last July and did the concrete work myself. I used a 6-bag mix with 5 percent entrained air and the concrete wasn't too soupy—not more than a 5-inch slump. It was a very hot day but my neighbor and I gave the surface a steel trowel finish before it had set up too much. I got some burlap bags from the local feed store and covered the surface immediately after finishing, then wet it down well and covered the whole thing with polyethylene sheets. I kept the burlap wet for a week. When I took it off there were several places on the surface where it looked as though the concrete still hadn't set. Some were as much as 6 inches across and others were only a couple of inches in diameter. The spots weren't more than ⅛-inch deep. The coarse aggregate was showing in these spots and yet most of the patio looked fine. What could have caused this?*

A. One possible cause for this problem is that some of the burlap bags you used were contaminated with an organic substance such as sugar that retards concrete setting. Because you gave the patio a steel trowel finish and may have finished it too early, it's also possible that blistering was the problem. Blisters occur when entrapped air or bleed water gets trapped in the concrete. However, since you said that it looked as though the concrete hadn't set in the affected areas, our first guess is the more likely answer.

Delayed setting of superplasticized concrete

Q. *What causes delayed setting of superplasticized concrete? We placed a 2-foot-thick slab on grade for a water treatment plant. Maximum air temperature during the pour was 65° F. We placed the concrete directly onto a polyethylene vapor barrier. At midnight, 10 hours after placing, the concrete was still liquid so the crew went home. Next morning the concrete had set. I'm investigating the problem but don't know exactly what to look for.*

A. Several factors probably combined to cause the problem. The first question to ask is whether there were any other admixtures in the concrete: water reducers, retarders, or fly ash. Combining a super with another admixture that retards setting can sometimes cause abnormally long setting times. For instance, concrete with only a water reducer might set in 4 hours. Concrete with only a superplasticizer might set in 4 hours. But concrete with both the water reducer and superplasticizer might take 6 hours or more to set.

What type of cement was used? For a water treatment plant, especially with the thick section, Type II or Type V cement may have been used. These types set slower than Type I or III cements.

Excessive bleeding may have slowed setting. Part of the stiffening process that permits finishing is caused by drying, part by cement hydration. Because the section was so thick and there was a vapor barrier beneath it, there was probably a lot of bleed water on the surface that prevented any drying.

Finally, temperature could be a factor. What was the temperature of the concrete as delivered? With an air temperature of 65° F, concrete may not have been much warmer. In thick sections it wouldn't be warmed much by the sun. And the temperature probably dropped even lower during the late afternoon and night. Lower temperatures increase setting time.

Find out if the combination of super, cement type, and other admixtures was used in trial batches under the same temperature conditions found at the jobsite. You may need to make some mix adjustments before placing more concrete.

Concrete sticking to curb forming machine

Q. *Concrete is sticking to the forms of our stringline curb forming machine and peeling off the newly formed curb as the machine moves on. Temperatures have been very warm, rising to about 90 to 95° F in midafternoon. What causes this problem, and what can we do about it?*

A. It's quite likely you have too much fine aggregate in the concrete mix. Try replacing some of it with coarse aggregate. Start by using 100 pounds more coarse aggregate and 100 pounds less sand and then adjust further as needed. Anything you can do to keep the concrete cool during mixing, transporting, and placing is also likely to help.

First batch in wall can be slightly wet

Q. *We seem to get more honeycombing at the bases of walls than higher up. Could this be because the first mixes were on the dry side, before we got the mixes adjusted?*

A. It could be. Concrete slump always varies slightly from batch to batch, and adjustments should be made on the basis of the slumps being measured as the day goes on. If any batch should be a little bit on the wet side it should be the first one. Don't let the first batch be too dry. Then make adjustments as needed.

Maybe right- and left-handed operators

Q. *Two operators did the chiseling to expose the aggregate on some areas of architectural concrete that we wanted to expose. It's pretty clear which operator worked on any particular area. How can we get uniform work when we need more than one operator?*

A. Any two operators are likely to produce differences in appearance because of differences in technique. This is particularly noticeable if you have a left-handed and a right-handed operator. One method is to put them to work on separate walls, or sections that are not adjacent. Another is to plan the work to deliberately work the differences into a pattern.

Is exposed aggregate concrete more durable?

Q. *At a regional shopping center in northern Illinois we noticed some good exposed-aggregate sidewalks which have been in place a number of years. Then we saw that a lot of the bands of plain concrete separating the exposed aggregate surfaces were in relatively poor condition. Is this happening elsewhere? If so, does it indicate superior durability for exposed aggregate work?*

A. Without knowing more about the construction history of the two concretes, we don't have an answer to this question.

Responses regarding durability of exposed aggregate concrete

The February issue showed a picture in Problem Clinic, page 150, of some exposed aggregate concrete flatwork in good condition adjacent to some badly scaled concrete of presumably the same age. The question was raised whether exposed aggregate concrete flatwork is generally more durable than troweled flatwork, other things being equal.

One contractor reader, Milton Collins, responded that better durability should be expected in the exposed aggregate concrete. Collins believes this is because the cement-sand matrix that remains exposed between the aggregate particles has not been subjected to the finishing operations that often, or usually, increase the water-cement ratio. The higher water-cement ratio weakens the surface and makes it less durable.

This may be true. Yet it does not explain why more material in the picture seems to have broken loose than just the top layer of cement-sand matrix. There appear to be some loosened aggregate particles as well. This suggests that a little farther down the concrete the scaling resistance was still deficient, even though it was probably not affected by the trowel.

Another reader, Edward M. Rizzo, of Preco Industries Ltd., replied that his company has test results on circular test specimens that were subjected to ASTM C 672 salt scaling exposures and then measured for weight loss after 50 cycles. Specimens whose surfaces had been exposed by retarders before the test lost only 60 to 80 percent as much weight as a control specimen. (The control specimen lost 80 milligrams per square centimeter of surface). Tests also showed that the weight loss of the exposed aggregate specimens could be greatly reduced by applying a good surface sealer.

Predicting rate of sulfate attack

Q. *When sulfate concentration is given in percent of soil or parts per million of water, is there a way of converting this value to pH so that the value could be used to predict potential sulfate attack on concrete?*

A. No. The pH of a solution (in this case sulfate in water) is a measure of its *acid* concentration, but it is really the concentration of *sulfate* ion which directly determines the rate of sulfate attack. For any particular sulfate salt the rate might be influenced by pH but not primarily determined by it.

As a matter of information it would be possible to convert the concentration of sulfate salt in water into pH, but only if one had a separate table for each sulfate salt. Furthermore, any pH value picked from such a table would be inaccurate if any other salts, alkalies or acids were present in the water.

To convert percent sulfate in soils to pH would be even more complicated and impractical as a means of predicting sulfate attack.

Corrosion

Corrosion of aluminum steps in manhole

Q. *During construction of a new sewage collection system it was found that the aluminum manhole steps embedded in the concrete manhole pipe had not been tar-coated before installation. All published articles we have found about galvanic effects of aluminum in concrete have said a protective coating should be applied to the aluminum to eliminate the possibility of corrosion.*

The manhole manufacturer claims that the aluminum steps do not come in contact with concrete containing calcium chloride but are in fact set in a proprietary patching compound. This contains, among other things, portland cement, sand, a synthetic organic polymer, an organic plasticizer and hydrocellulose thickener and no calcium chloride. Will there be a corrosion problem in the future?

A. It is not entirely clear what all the ingredients are—that is, whether the synthetic organic polymer is a latex binding agent or a water-reducing agent of some kind, and what the organic plasticizer is. Nevertheless, we doubt that the patching compound would attack aluminum.

On this basis we would not suggest having the aluminum steps removed, dipped and reset. You might be justified, however, in asking for a brief warranty, such as a year or two, during which you could watch the performance of the steps and have some tests made if you see anything suspicious, such as cracking around the steps or buildup of salts at the point where the aluminum steps enter the concrete.

Sound the alarm

Q. *As a member of our city building department I have been asked to solicit comments from the private sector regarding the attached article. [The inquirer enclosed a copy of an article by William H. Hartt and Philip E. Martin, ''The Environmental Deterioration of Concrete Construction: An Emerging Problem in South Florida,'' Florida Environmental and Urban Issues, January 1981, pages 10-14.]*

A. To make clear our comments on this article we should point out that the ''emerging problem'' in the title is not some new disease of concrete. The emerging problem, according to the authors, is that insufficient attention is being paid by designers and constructors to the known methods and code requirements for protecting reinforced concrete against chloride attack in coastal areas. As a result, according to the authors, ''it is quite possible that the damage development clock is ticking and that it is simply a matter of time before concrete deterioration becomes more evident.''

Cracked structural column, with spalls, of coastal high-rise building.

Problem Clinic does not presume to be informed about current design and construction practices in South Florida. We do agree, however, with the authors' discussion of the mechanism of chloride attack on concrete, the means of preventing it and the consequences of not designing and constructing in ways that prevent chloride attack in coastal areas. For this reason we summarize here the discussion presented by Hartt and Martin in the article cited. The authors are professors of Ocean Engineering at Florida Atlantic Univerisity. Although the article was written about what they perceive to be an emerging problem in South Florida it is possible that the same problem may exist in a number of other areas.

The three photos illustrate examples of deterioration. The first is the concrete piling of a 20-year-old bridge in coastal waters. The piling is badly cracked and spalled. The second is a cracked structural column of a coastal high-rise building. The third is a balcony railing of a 12-year-old coastal high-rise condominium.

Cracked and spalled piling of 20-year-old bridge in coastal waters.

This kind of cracking and spalling invariably results from corrosion of reinforcing steel, causing accumulation of corrosion products (rust) on the steel surface. Continuing deposition of rust creates tremendous internal tensile stresses within the concrete that result in cracks that parallel the length of the embedded metal. Unfortunately, since

Cracked balcony railing of 12-year-old coastal high-rise condominium.

each crack begins from the inside out, damage is usually not seen until the crack reaches the concrete surface. By this time deterioration is already very advanced.

In most exposures steel does not corrode in concrete because of the alkaline nature of the concrete. However, chlorides from the environment may slowly permeate concrete and reach the metal surface. In coastal locations chloride in sea spray can lead to the kind of disruption shown in the photos.

Following are the factors that have the greatest influence on the amount of corrosion damage:

- chloride concentration in the environment
- chloride concentration in the aggregate
- permeability of concrete
- thickness of concrete cover over reinforcing steel

The greater the amount of salt spray or other source of chloride the greater the intensity of the attack. The presence of chlorides in aggregates has often been overlooked; aggregates taken from coastal sites below sea level—or even slightly above sea level—frequently contain chlorides. It is thought that one pound of chloride per cubic yard of fresh concrete is enough to cause corrosion of embedded steel. Permeability of concrete usually decreases as the strength goes up. For this reason it is generally desirable to use a high but practical concrete strength to minimize penetration by chloride solutions and consequent corrosion. Cover thickness affects the rate at which chlorides are able to reach the metal-concrete interface.

The amount of cover over any steel reinforcement is specified in building codes. In general 1½ to 2 inches of cover are required over steel in concrete that is exposed to weather. Most codes state that in corrosive atmospheres or in concrete with high porosity cover should be increased.

Unfortunately architects and construction engineers, according to the authors, are generally unfamiliar with the susceptibility of concrete to cracking and spalling in coastal environments, and the requirement for extra cover is often overlooked. Even worse, in South Florida it has become common to specify ¾ inch of cover over steel. In recognition of this reality local boards of rules and appeals are said to be considering changes in requirements so that the specified concrete cover agrees with past and current local practice. The authors believe "that any code revision which reduces the amount of concrete cover over reinforcing steel has potentially disastrous consequences. If anything, the specified cover for concrete exposed to weather should be increased, not decreased." Although this change would add considerably to initial costs it could save tremendous costs created by the long-term consequences of inadequate cover. The authors argue that cover requirements should not be reduced below those presently specified in codes and in fact "professionals in the construction field must develop a greater appreciation for the code statement that increased concrete cover should be provided in corrosive atmospheres."

The authors also point out a series of preventive measures that can be used to minimize or prevent concrete cracking from corrosion of embedded steel:

- use of sound design and construction practices
- use of proper materials and techniques of material preparation
- use of coated reinforcing steel
- application of protective coatings or penetrating sealers to the external concrete surface
- use of waterproof membranes
- polymer impregnation of concrete
- use of inhibitors
- cathodic protection

The authors point out that the first two of these are particularly important in order to achieve concrete with high strength and low permeability and to provide adequate cover for the steel. The next four measures are ways of making reinforced concrete more resistant to movement of chloride solutions within the concrete. Each of these methods may be effective to some degree. Inhibitors such as calcium nitrite have been shown in recent research to prolong the time before corrosion can begin. Cathodic protection has been applied to a number of bridge decks in the United States and Canada and also to concrete offshore petroleum production platforms in European waters. Its application to high rise buildings and other structures with more complex geometry would probably be impractical.

Since this type of corrosion occurs at an ever-increasing rate, the authors say that current damage may be a preview of what may be a much more serious problem 10 or 20 years from now. They point out that "the problem is compounded by several factors, the first of which is the difficulty associated with providing adequate repair and also uncertainties regarding the effectiveness of various preventive techniques. Most significant, however, is the general lack of experience by architects, engineers and construction personnel in dealing with this problem. This is compounded further by the fact that construction codes dealing with concrete do not adequately emphasize the problems that can arise in locations where chloride penetrations must be anticipated, and persons charged with overseeing these requirements apparently do not realize the importance of the reinforcing-steel-corrosion concrete-cracking problem." They recommend that engineers, architects and construction personnel make themselves more aware of this problem and become familiar with the techniques for reducing and preventing the damage. They also recommend that existing construction practices as specified in various codes should be modified to deal with the problem.

A better deicer coming?

Q. *What's this we hear about calcium magnesium acetate as a deicer that doesn't corrode steel? Is somebody making it, and if so where do we get it?*

A. It's an experimental product whose manufacture is still being developed and would be pretty expensive if made in the small quantities and by the methods used so far—perhaps 18 to 19 cents a pound exclusive of distribution costs. It is reported to be noncorrosive to steel, and it's made by reacting dolomite (mainly a mixture of calcium carbonate and magnesium carbonate) with acetic acid. The Federal Highway Administration, the Michigan Department of Transportation and others are taking an active interest in the manufacture of the product and in studying how well it performs as a deicer. Only experimental quantities have been made so far. We hope to keep you posted about any new developments.

Protecting lead in concrete

Q. *Is there any experience with lead embedded in concrete? A lead-sheathed cable is intended to pass through a 10-foot length of concrete slab-on-grade. Will this cause any problems with the concrete?*

A. If there are no other metals in the concrete it shouldn't harm the concrete. As long as the slab stays moist, however, the calcium hydroxide in the concrete can attack the lead and within a few years the lead sheath could be disintegrated. The sheath could be protected from the concrete with a thick bituminous coating or by enclosing it in a polyethylene or other alkali-resistant plastic sleeve.

Slab disrupted over conduit

Q. *The surface of a 12-year-old slab has heaved above the embedded electrical conduit. Wherever the concrete has been disrupted enough to expose the conduit to view, a white-colored deposit can be seen. This looks much like the deposit that develops on the post of an automobile battery. What has caused the heaving? We think calcium chloride was used in the concrete mix and wonder if that could have had anything to do with the trouble.*

A. What you describe is typical of galvanic corrosion. This type of corrosion can occur wherever two dissimilar metals within the slab are in contact with one another. Salts of the conduit metal (perhaps zinc salts or aluminum salts) form on the metal surface, creating expansive forces that disrupt the concrete. The corrosion is likely to be even more severe if the concrete contains calcium chloride.

Corrosion potentials in field may surprise you

Q. *Why and how does a copper water line encased in concrete corrode if it is touching a piece of reinforcing steel? I have looked at tables of corrosion potential and they show copper to be more noble than steel. This says to me that the steel should corrode instead of the copper. What gives? Does it have anything to do with the water flowing through the pipe? Are there any studies or reports covering this phenomenon?*

A. The subject is more complicated than the usual table of electromotive series potentials might lead you to believe. Such tables are usually based on potentials measured in specified salt solutions of given concentrations at one temperature. If you change the kind or concentration of solution, or change the temperature or the composition of the alloy, everything may change. The relative positions of the two metals may widen or narrow or they may even reverse. Thus if Metal A and Metal B are in contact with one another and with some specific electrolyte, Metal A may corrode. But if you change the concentration of the electrolyte, the composition of the electrolyte, or the temperature, or if you use a different alloy of Metal A or Metal B or both, it is possible that Metal B will corrode instead of Metal A.

Some books you might like to look into are:

Corrosion of Metals in Concrete, Publication SP-49, American Concrete Institute, Box 19150, Detroit, Michigan 48219, 142 pages. Published 1975. A collection of 10 state-of-the-art reports on corrosion mechanisms, field procedures and research.

Chloride Corrosion of Steel in Concrete, Special Technical Publication 629, ASTM, 1916 Race Street, Philadelphia, Pennsylvania 19103, 193 pages. Published 1977. A collection of 17 papers on research and field practice in controlling corrosion of steel in concrete.

Corrosion of Reinforcing Steel in Concrete, Special Technical Publication 713, ASTM, 1916 Race Street, Philadelphia, Pennsylvania 19103, 204 pages. Published 1980. A collection of 11 articles on the mechanisms of corrosion of steel in concrete and methods of evaluation and control.

Concern over cover after sandblasting

Q. *We are concerned with the maintenance of a cast-in-place school building built in our town in 1937—one of four local schools built by the Works Progress Administration (WPA). In its lifetime it has had several coats of paint. Finally it has become necessary to* *remove paint. We planned to remove it by sandblasting but we were told that this would remove a surface layer of cement and sand, making the reinforcing steel vulnerable to corrosion. Do you think it is safe to sandblast?*

A. The amount of concrete that you remove by sandblasting the surface to remove old paint is not likely to leave the concrete vulnerable to corrosion. It is true that the thickness of concrete covering the steel is a determining factor in whether or not the steel will corrode. Consequently there are standard specifications giving the minimum amount of cover required for various kinds of service. The required cover depends on the type of exposure and the type and size of the steel reinforcement. You can find the standard requirements in ACI 318, "Building Code Requirements for Reinforced Concrete."

Removing up to 1/8 inch of concrete by sandblasting should not affect the amount of concrete cover substantially. Of course if the building did not originally have the required concrete cover or if you are still concerned about having enough corrosion protection after sandblasting, you could apply a sealer or new paint coating.

Preventing rust stains

Q. *We were very pleased with our architectural concrete office building until ugly rust stains appeared on several surfaces. Did the contractor make a mistake, or is it unavoidable? What can be done?*

A. There are several possible causes. Occasionally iron sulfide particles in the aggregate are at fault. Loose iron or steel particles left in inadequately cleaned forms can also cause trouble, as can reinforcing bars too close to the concrete surface. Bar supports that were't protected against corrosion can produce small rust stains in a recognizable pattern. At times, a reinforcing bar is incorrectly positioned too close to the concrete surface and starts to rust.

Removal of the stains depends on where and how deep they are. Both chemical and mechanical means have been used. Detailed instructions for removal of iron rust stains are given in the August 1986 issue of *Concrete Construction*, page 742.

On future work, consider preventive measures such as careful form cleanup and use of enough noncorroding bar supports to keep reinforcement the specified distance away from the concrete surface. Plastic-tipped or stainless steel bar chairs are available. All-plastic spacers like those shown in the photograph also work well.

Plastic sideform spacers can maintain the desired distance between a vertical form and reinforcing bars. The spacers snap onto the bars and are usually pierced with large openings to ease flow of concrete around the steel. Spacer at left, designed for use with foamed plastic liners, has feet that punch through the liner to the form backing.

Cracks in the curing membrane

Q. *We completed a floor pour and were surprised to find fine pattern cracks in the surface the next day. The mix wasn't particularly rich, with a cement content of 550 pounds per cubic yard. We carefully limited slump to 3 inches and didn't overtrowel the surface. Because the weather was hot and dry, we applied a curing compound as soon as the bleedwater sheen had left the surface. However, the cracks were quite visible the next day and the owner was upset. We made six pours on the job and had this problem with only one of them. What did we do wrong?*

A. There's a possibility that the cracks were only in the curing membrane and not in the concrete. The American Concrete Institute's "Standard Practice for Curing Concrete" (ACI 308-81) warns that map cracking of a membrane film can be caused by applying curing compound to a dry surface before bleeding has stopped. This is most likely to happen when the evaporation rate exceeds 0.2 pound of water per square foot per hour. When this happens, water evaporates faster than it can rise to the surface through bleeding. If a membrane-forming curing compound is applied to a dry-appearing surface, evaporation will be temporarily stopped but bleeding may continue. This floats the membrane film which cracks when the water later evaporates.

To find out if cracks are in the membrane instead of the concrete surface, you might try rubbing a small area with additional liquid curing compound. If the cracks disappear you'll know they were in the membrane. Then apply more curing compound over the cracked area to ensure that internal moisture is retained.

Cracking of new topping

Q. *We contracted to remove and replace a cracked and broken topping in a 6 x 40-foot area of the second floor of a two-story concrete garage. The structural steel of this building, built about 50 years ago, is buried in the concrete. Although the owner thought that the old topping was a few inches thick, it turned out to vary from only $^1/_2$ to $1^1/_2$ inches thick, and part of this variation occurred because the surface itself was far from flat. Because the topping was so thin I had to give up the idea of using mesh or even coarse aggregate, and we ended up using a 4000-psi air-entrained mix of cement, sand, and water. We kept the base concrete damp but not moist, kept the slump low, placed concrete about 3 P.M. gave it a steel-troweled finish and cured it with a membrane curing compound that we use on state work.*

A week later the owner noticed several cracks. By age 6 weeks random cracking is extensive, cracks are wider, and pieces of mortar can be broken out without much effort. The only thing I know that we may have done wrong was to allow cars to be parked on the repaired area on the third day after we placed the concrete. What should we have done or what should we do?

A. We suspect that the trouble lies in the relatively high drying shrinkage that most 4000-psi mortars would be expected to have. Normally you could solve that problem by simply applying a topping of concrete, instead of mortar, so that there would be coarse aggregate present to reduce the drying shrinkage. But that would mean either (a) using a thicker topping, which would add to the elevation and to the dead load, or (b) cutting into the structural steel below the topping. Probably neither of these would be allowed.

The following suggestions for repair are based on the assumption that you would completely remove the existing topping and then roughen and clean the surface adequately to accept and bond to a new topping. No matter what topping you use, it might be wise to ensure a good bond by scrubbing fresh 1:1 cement-sand grout into the substrate just ahead of where you're placing the topping.

You might consider using a mix that incorporates polypropylene fibers to minimize crack propagation. There are also companies that make packaged repair materials that contain fibers in the mix.

Perhaps you should also investigate acrylic concretes. These are made by some of the companies listed under "Concrete Polymer Materials" in the Buyers' Guide (yellow pages) of our December issue and are discussed in the article "Acrylic Concrete for Fast Repair," page 881 of our October 1984 issue.

Run joint to apex of any boxout in flatwork

Q. *In a parking lot we constructed we made sure that wherever we had a planter we ran our joints to the planter. This left four returns in the adjacent panels, and cracks formed at a good many of the corners, as in the drawing. What's the best solution?*

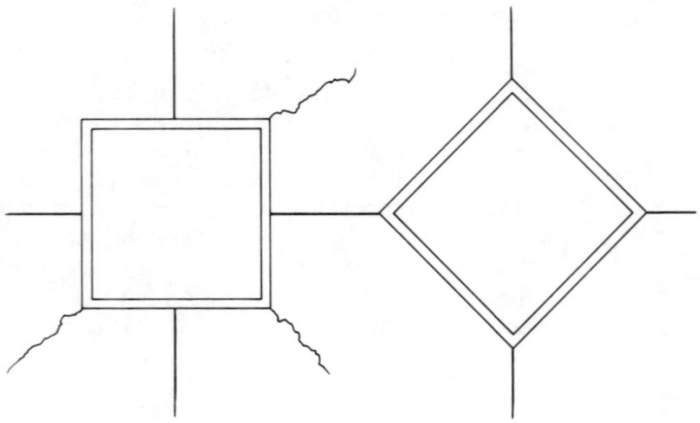

A. The planters should be rotated 45 degrees so the joints meet the apexes of the square, as in the second drawing. The same would be true if the areas concerned were being used for drains or any other purpose. The principle is widely advocated and understood for boxouts around columns to be isolated from floors on ground, but often it is forgotten for other boxouts.

Plastic shrinkage cracking

Q. *On a spring day we were placing slabs that cracked open before we could finish them. The ambient high temperature was 40° F. We then asked the ready mixed concrete company to send concrete containing retarders and they tried two different retarders in different proportions. The concrete then began hardening on the surface even faster.*

Now the only admixture we ask for is an air entraining agent, but we still get cracking. One day this spring, a slab that took us four hours to place from start to finish cracked all over. There was some sun and wind, and the temperature was 60° F. The concrete was a 5½-bag mix with a 5½-inch slump placed on a rock fill. We have to use the 5½-inch slump because of the high absorption of the lava rock and poor sand used for the fill.

We have less trouble with 6-inch floors than 4-inch, but 6-inch floors crack also. The cement is not excessively hot. Will too much water reducer cause trouble?

A. The problem seems to be what is known as "plastic cracking" or "plastic shrinkage cracking." This can occur under any conditions where evaporation is rapid—whether hot, dry weather or any kind of windy weather (hot or cold). What often happens is the surface dries rapidly and shrinks enough to cause many cracks, usually parallel to one another, before the concrete has been finished.

Because plastic shrinkage cracking is related to evaporation rather than rate of chemical hardening, the use of retarders to control it is ineffective. The air content should not affect the problem. The effect of water reducers may be minor.

The solution is to do everything possible to protect against rapid evaporation. Some measures that can be taken are described in two articles from *Concrete Construction*: "How Cracking is Controlled," November 1973, page 517 and "Finishing Problems and Surface Defects in Flatwork," April 1979, page 247.

Cracking over rebars in flatwork

Q. *We do finish work for a pool contractor who insists on using #4 rebar in all his flatwork. We get hairline cracks above the rebar locations except where concrete is 6 inches thick. We have suggested that he use welded wire mesh but he refuses. What can we do to stop the cracks?*

A. Your suggestion of using wire mesh (which has a smaller diameter than #4 bars) and your observation that cracking doesn't occur when the steel is deeper in the concrete agree with the recommendation of the ACI 302, Section 2.2.2, which appears in *Concrete International,* June 1980.

The first paragraph of that section says:

> "Plastic settlement cracking over rebars is caused by inadequate concrete cover over steel, larger rebar size and/or higher slump concrete or a combination of these. To help avoid these problems, concrete cover should be at least 2 inches (50 millimeters) where possible. Other measures to avoid this are smaller bar size, low slump concrete (2 inches or 50 millimeters) and/or revibration of concrete."

This recommendation is based on data presented by Fadh. H. Dakil, Philip D. Cady and Roger E. Carrier in the article "Cracking of Fresh Concrete as Related to Reinforcement," *ACI Journal,* August 1975, pages 421-428.

ACI 302 also says (Section 11.1.3):

> "Some cracking may occur before the concrete has hardened. This may complicate the finishing operations considerably. Some examples are: . . . (2) cracking from settlement of concrete around reinforcing bars or other embedments."

Preventing cracks in superflat floors

Q. *Is it true that superflat floors are built without transverse joints and that they are expected to crack? Is it possible to build a superflat floor that doesn't crack, with or without transverse control joints? We need a floor that is crack-free for reasons of sanitation.*

A. Joints are omitted from superflat floors because the joints would tend to become high ridges in the floors. A possible solution to your problem is to build a post-tensioned superflat floor. This could be post-tensioned in either one or two directions, depending on requirements. George Garber reported on construction of both one-way and two-way post-tensioned floors in the May 1983 issue of *Concrete Construction*, page 396.

More on preventing cracks in superflat floors

After this question was addressed in Problem Clinic in our July 1984 issue, page 679, a reader responded that our answer was incomplete, especially on the subject of control joints, and offered the following information:

Another solution is to use shrinkage-compensating cement concrete, as discussed in the following articles in *Concrete Construction*:

> "Shrinkage-Compensating Cement Concrete," February 1976, page 63.
>
> "Good Warehouse Floors Don't Just Happen," September 1977, page 493.
>
> "Shrinkage-Compensating Concrete for Industrial Floors on Ground," February 1979, page 110.
>
> "How Does Shrinkage-Compensating Cement Concrete Help Post-Tensioning?" April 1983, page 331.

Anyone interested in this subject might also want to study ACI 223, "Standard Practice for the Use of Shrinkage-Compensating Concrete,"

Although some engineers feel that shrinkage-compensating cement concrete is a competitor for post-tensioned concrete, actually the two complement each other when combined on a project.

Cracked driveway corner

Q. *A customer complains that the corner of his driveway next to the street cracked when a heavy truck made a delivery. This was a 5-inch slab. Should the driveway have been doweled to the street pavement, or what is the best construction method to prevent this from happening?*

A. It has been recommended that the driveway slab be thickened 25 percent at the junction—that is, at the curb line if there is one, or at the pavement edge. The thickened end should be tapered back for a distance of 5 feet. If there is a sidewalk parallel to the street that intercepts the driveway, the driveway should be thickened on both sides of the sidewalk.

Preventing cracking of driveway corner

Q. *In the August 1984 issue, page 768, of* Concrete Construction, *a recommendation was given on how to prevent cracking from truck loads where residential driveway slabs meet sidewalks or street pavements. It was recommended that driveways be thickened 25 percent at their junctions with street or sidewalk pavements and tapered back to normal thickness over a distance of 5 feet. Theodore Hemleben, of Hemleben Construction Company, Wapakoneta, Ohio writes to suggest a different method. He writes:*

A. Try this answer. Use one or two pieces of 1/2-inch reinforcing bar, 2 to 3 feet long in the corners as shown in the drawing. With steel used in this way the con-

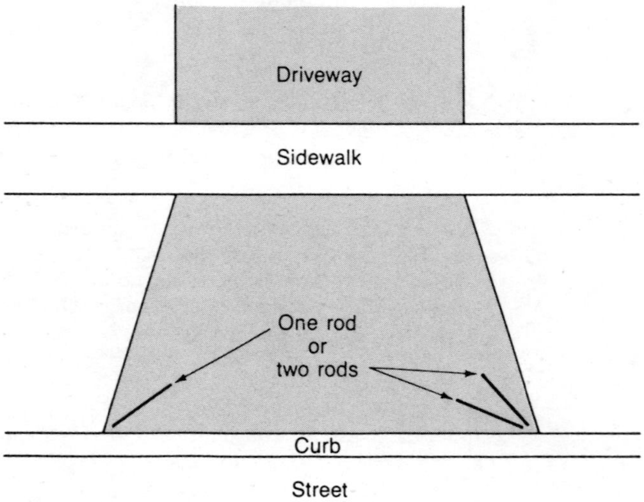

crete won't crack from loading even if the driveway is only 4 inches thick. By the method recommended in the August issue the concrete could be 10 inches thick and still crack. Note also that the sidewalk should be 5 inches thick if, as Problem Clinic seemed to say, it is to intersect a 5-inch-thick section of driveway.

Controlling cracking of driveway corner

In Problem Clinic in the August 1984 issue, page 768, we gave a method for preventing cracking of driveway corners and in the September 1985 issue, page 793, we gave reader Ted Hemleben's alternative method (using rebars). William F. Klosterman, city engineer of Middletown, Ohio has sent the following suggestion.

A third alternative, which we have used for 15 years in Middletown, Ohio, is to allow the crack but to *control* it by merely striking a joint as shown in the drawing. The joint should form an isosceles triangle whose two equal sides are each 2 feet long.

ACI Committee 302 says that tooled joints to control cracking must be cut 1/4 the thickness of the slab.

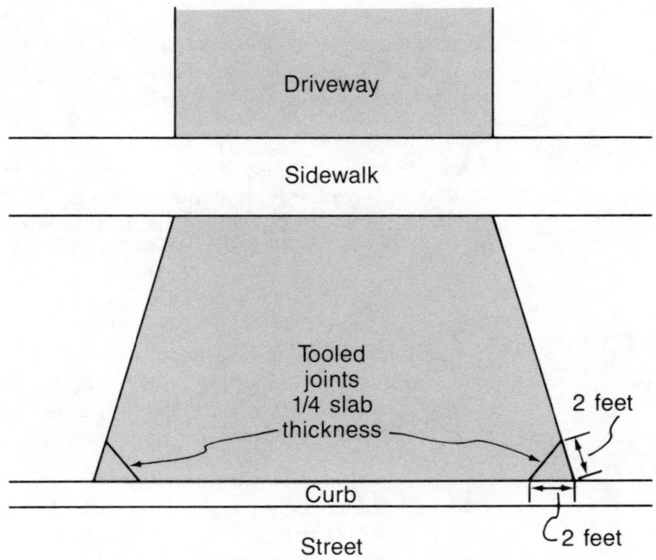

Cracking over rebar chair

Q. *We pour parking blocks with leftover concrete using a plastic rebar chair to hold the 1/2-inch rebar in place. We almost always observe a crack where the plastic chair is placed in the form. The chair has to rest on the bottom of the form. What causes the cracking? Is there a remedy?*

A. We are not sure what causes the cracking unless it is the weakening of the section by the plastic chair. If that is the cause you might try switching to the type of chair that consists of a small concrete cube with wires protruding from the top. The rebar is placed on the cube and the wires are then twisted to hold the rebar in place. The concrete cube, of course, becomes a permanent part of the parking block.

If you cannot buy such chairs in your area you could probably make your own.

Preventing cracking in parking blocks

Robert J. Ilukowicz, Director of Corporate/New Product Development at Preco Industries Ltd., has additional information on the Problem Clinic item "Cracking over Rebar Chair," published on page 702 of the September 1980 issue.

Because of the difference in the coefficient of expansion of plastic and concrete it is sometimes too easy to blame the plastic rebar chair being used. Shrinkage of concrete during the curing process or corrosion of metal reinforcement may induce cracking near the rebar chairs as well.

It has been our experience over the last decade that with a properly designed plastic rebar chair many of the potential shortcomings are more theoretical than real. I am particularly referring to the perforations intentionally made in the piece to ensure maximum bond with the concrete.

Since we have patented a rebar chair for a bumper curb, a viable alternative can be offered. Your reader should also consider the fact that his concrete mix may be too rich as well as consider covering his finished product with a polyethylene blanket to minimize the moisture loss.

Cracked tees in cooler

Q. *There has been serious spalling in the stems of precast tees of the ceiling in a cattle carcass cooler. The tees are 18 inches high with rebars located 3 inches, 5 inches and 7 inches from the base. Carcasses are run into the cooler on tracks suspended from the tees. The cooler operates at a freezing temperature of 28° F but while the cooler is being loaded* *with new carcasses the temperature rises to 40° F. The cooler has been in use for 8 years and, because of the carcasses, the inside atmosphere has had a high humidity.*

Recently, in order to bring the temperature down faster, carbon dioxide gas was injected through the doorway. This rapid cooling was done a number of times. Some tees near the doorway have now shattered in long slim fractures resembling those of overloaded columns. It is possible that the tees were overloaded with too many carcasses. Some of the fractures are located as high as the highest rebar. What is the likely cause of the fractures?

A. The most likely cause is that the carbon dioxide gas rapidly froze the concrete which had probably become thoroughly saturated. The very low temperature of the carbon dioxide gas perhaps caused the beams to freeze at the bottom much faster than at the top causing the high stresses.

Cracking at columns in outdoor flatwork

Q. *Why do we see so many cracks in flatwork around small columns? The problem seems epidemic in shopping centers. For example, in one shopping center in our area the quality of the concrete was generally good except at columns. Unfortunately, there were many columns.*

A. There is an old, well-established rule that seems to be widely violated: *Always* isolate columns from the flatwork by means of joints. There should always be a joint close to the column and completely surrounding it. A common method is to wrap the base of the column with a strip of asphalt-impregnated fiber 1/4 inch thick and 4 or 6 inches high. Where the two ends meet it can be held in place by tape running circumferentially.

But that is not all. There must be contraction joints running to the column. Difficulty often arises because the contraction joint pattern has been established without reference to the column location. Then, even if this pattern is later adjusted, the joints may not be located and installed in a way that makes them function properly. The contraction joint pattern should cause every column to fall symmetrically along the center line of a contraction joint. These contraction joints should be real working joints, not just dummy joints. This means they should be made 1/4 or 1/3 the thickness of the concrete.

Cracks at re-entrant corners

Q. *In our part of Florida, concrete slab floors are used for residential and commercial buildings, whether the walls are concrete block, wood frame or metal. Typically cracks develop at the locations shown (Figure 1). How can these be prevented? Wire mesh and a*

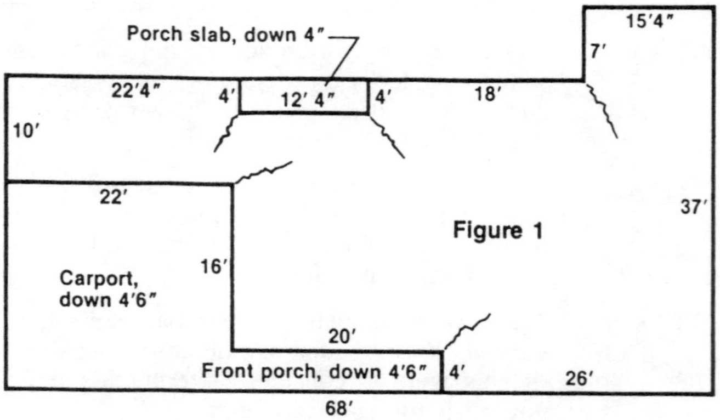

Typical slab showing probable crack locations.

plastic vapor barrier are used, but no other bars are implanted. Will the new polypropylene fibers in the concrete be of help? I am not a concrete contractor but a builder of residences and small commercial buildings. Other builders here would be happy to know how to prevent this cracking.

A. This kind of cracking in concrete commonly occurs at re-entrant corners, that is, corners that are directed inward rather than outward. We do not think using polypropylene fibers in the concrete would be likely to prevent the cracking because of the high stress concentrations at these corners. The preferred method of handling this problem in floors is to install joints that break up the slab into panels so that there are no re-entrant corners *within* panels. In the example you provided, this could be done in a way such as that shown in Figure 2, but there also could be other ways of dividing the slab.

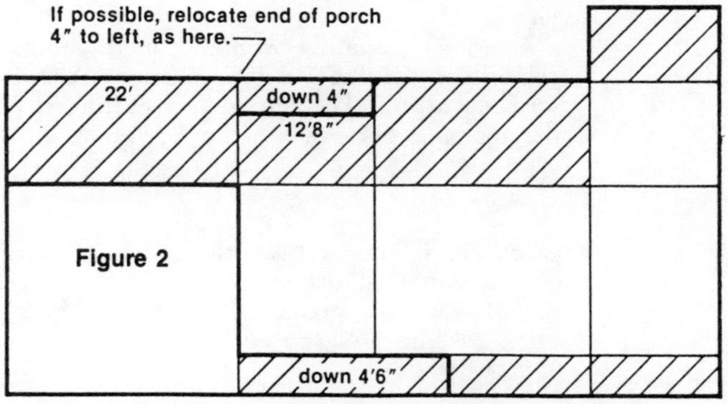

Install joints to eliminate re-entrant corners. However, this leaves the problem of many panels (crosshatched) with lengths more than 1.5 times their widths.

If possible, however, it would help if the architect could redesign the slab layout on some modular system that would allow the floor to be divided into approximately square panels no more than 15 feet on a side and without re-entrant corners. The existing layout makes it impossible to do this without creating a number of panels whose ratios of length to width are more than the recommended maximum limit of 1.5. Figure 2 shows six such areas in the floor itself plus two in the porches. There is no way to avoid overly long panels with the existing slab outline.

Another way to solve the problem is to insert reinforcing bars in the positions shown in Figure 3. Bars should be Number 4 bars or larger and they should be supported

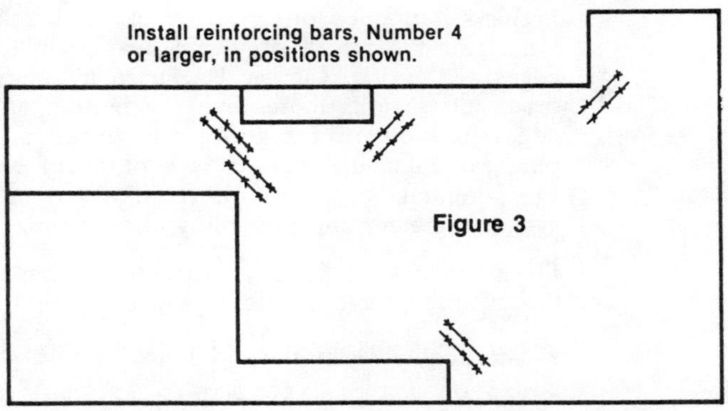

on chairs so that they will be located at mid-height of the slab. This should replace the large cracks with numerous unobjectionable tiny cracks too small to notice.

Curling

Thickened edges & curling

Q. *Curling is a big problem in our part of the country and we are concerned about ways to prevent it. Would thickened edges of the slab reduce or prevent curling, and if so, how much would they have to be thickened?*

A. Thickened edges would reduce or prevent curling, but because they provide additional subgrade restraint they could readily cause cracking elsewhere in the slab. It doesn't seem like a trouble-free solution.

Thick slabs curl less

Q. *We've had many recommendations on how to minimize curling of concrete floors—mostly recommendations about mix proportions, improving the possibility of moisture escaping from the bottom of the slab, and shortening the joint spacing. Would thicker slabs curl less?*

A. They certainly would. Robert Tobin, Regional Structural Engineer in the Los Angeles office of the Portland Cement Association, has developed a formula that describes curling. The amount of curling depends on the difference in shrinkage between the top and bottom, s_c, the length of the slab, l, and the thickness, t:

$$\triangle = \frac{s_c \, l^2}{8 \, t}$$

where $\triangle$ is the vertical movement of an end. Since the l term is squared, the length of the slab (distance between joints) is by far the most important factor. The longer the distance the greater the curling. Thickness affects curling inversely; the greater the thickness the less the curling.

Curled topping

Q. *We contracted to pour a 22x38-foot slab in an existing basement. Because of headroom, it was decided the slab would be an 8-bag grout mix only 2 inches thick. We used #3 rebars 30 inches on center the long way and poured over an existing concrete floor. The existing floor was badly cracked, but reasonably true and showed no signs of ever having settled or shifted.*

Within two months the new slab developed many shrinkage cracks, but the puzzling thing was one crack that ran generally lengthwise approximately in the center. About 15 feet of this crack was displaced vertically nearly one-quarter of an inch. Why would one side of the new slab raise like this in the center? The existing slab was probably at least 25 years old. The new slab was poured in warm weather, about mid-May, and because there was a building over the basement, curing conditions were quite good.

A. The vertical displacement is probably due to curling of one side of the topping. Perhaps it bonded on one side of the crack but not on the other. If you tap the topping near the raised edge and hear a hollow sound, it's a good indication that the edges are curled. Curling is caused by uneven shrinkage of slab concrete. The top dries and shrinks more than the bottom of the slab or topping. This tends to lift the edges. If there isn't a good bond between the topping and the base concrete, the topping can raise up near a crack or at an edge. A sand-cement grout has a high shrinkage potential because of the high water and cement content. The greater the shrinkage, the greater the possibility that curling will occur.

Eugene Boeke of Beers Construction Company in Atlanta gives this advice for avoiding the curling problem with toppings:

- Thoroughly clean the old floor. Waterblasting or sandblasting will work.
- Wet the old floor for at least 12 hours by spraying it and covering it with wet burlap.
- Just before applying the topping, scrub the floor with a slurry of neat cement, bonding agent, and water. Use a stiff bristle broom and lots of elbow grease.
- Use the largest possible aggregate size for the topping mix, the lowest possible slump, and the lowest strength mix you can still work and finish. Boeke uses a 2500-or 3000-psi concrete. If the strength is less than that, his finishers have trouble closing the surface. Low strength, low slump, and large aggregate reduce the amount of water and cement in the mix. This reduces shrinkage.
- Cut joints on a grid pattern after the topping is placed. Make them about 3/4 inch deep and 8 to 12 feet apart. These can later be closed by the finishers but will still give a plane of weakness and prevent major curling. Some people saw the

joints after 12 hours.

- Apply a good quality curing compound *immediately* after finishing is completed. Use twice the recommended application rate.

Boeke's crews have also had good results with a system they use for tying down the corners of a day's topping pour. They shoot three studs into the substrate using a triangular pattern as shown in the drawing. The studs are about 8 inches apart and 6

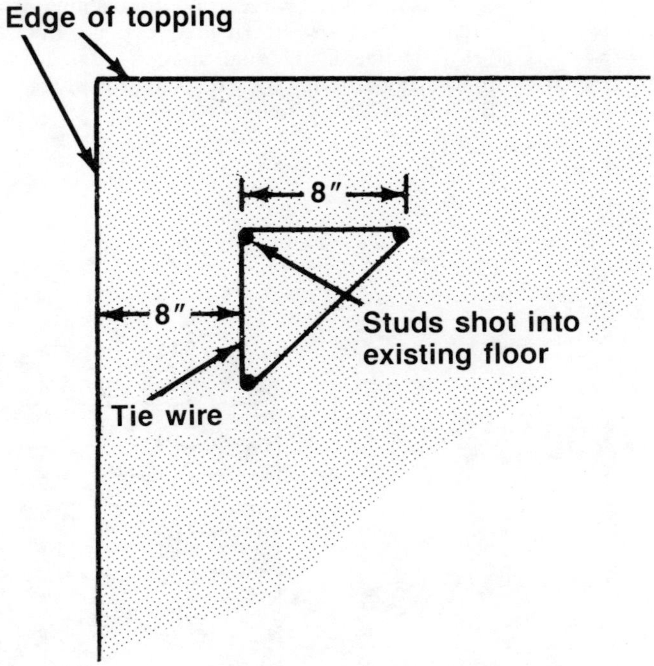

to 8 inches away from the edge at each corner. They are wrapped and laced with tie wire before the topping concrete is placed. This helps hold down the corners. If reinforcement is needed only for crack control, Boeke generally uses polypropylene fibers in the topping rather than small diameter rebars. If the reinforcement is for structural purposes, an engineer should determine the bar size and spacing required.

Preventing curling of slabs

Q. *Is there anything one can do, short of putting a slip joint between a slab and the masonry bearing wall that it rests on, to prevent breaking the bond to the mortar joint between them? We built a 12x15-foot vault inside a building for fire protection of documents. The top slab has now curled up about 1/4 inch at the corners, leaving a gap. We expect to tuckpoint the joint, but it will not look as neat as originally.*

Slab curling doesn't appear to be a problem with slabs on grade. Is this because it is not noticeable or because the subgrade controls the rate of curing better? Is there such a thing as curling in footings? I would appreciate any information or references on curling and concrete shrinkage.

A. We haven't heard of any problem with curling of footings, but curling *is* fairly common in slabs on grade. It is not always noticeable, but it can be troublesome in slabs where furniture or high shelving rests partly on a curled corner or where truck traffic crosses a joint between curled slabs. The main cause is greater shrinkage on the top surface than on the bottom. Anything that can be done to keep the moisture content of the slab more nearly uniform from top to bottom will help minimize curling. In Section 11.11 of ACI 302.1R-80, "Guide for Concrete Floor and Slab Construction," there is a fairly complete list of measures that can be taken to minimize curling, some of which would apply to your problem. This useful publication of 46 pages is available from the American Concrete Institute, Box 19150, Detroit, Michigan 48219.

Deflection can curl slabs too

In Problem Clinic for June, page 571, we answered a question about upward deflections of slab corners. A 1/4-inch upward deflection had been observed in the corners of a 12x15-foot concrete slab constructed on masonry walls. Both the question and answer were focused on slab curling caused by differential shrinkage. But Bruce A. Suprenant, of the Department of Civil Engineering of the University of South Florida, in Tampa, says the curling should be expected from load considerations without regard to shrinkage. He provided the following discussion:

The deflection is probably caused by the slab acting as a plate. For example, take an 8 1/2x11-inch sheet of cardboard backing from a writing tablet and cut it to 8 1/2 inches square (don't use corrugated paper, which is stiffer in one direction than the other). Set four books of equal size on edge to support the cardboard. Using your hand, load the cardboard downward as uniformly as possible. You should observe an upward deflection at the corners.

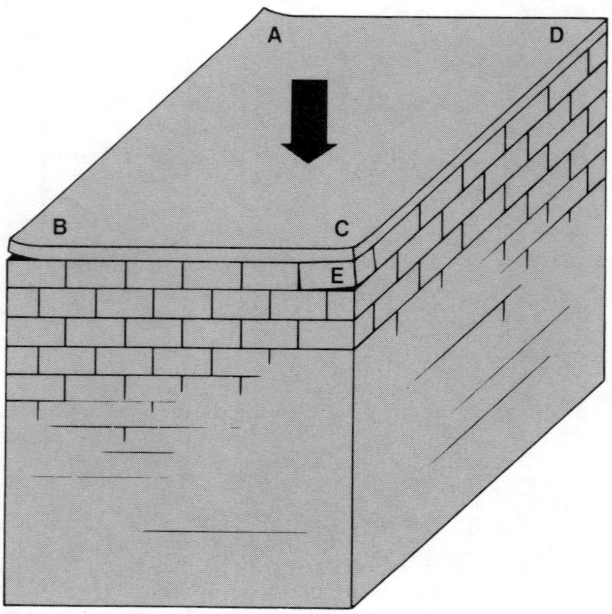

Figure 1. *Plate action can cause upward curling of slab supported by masonry walls. Corners at A and B represent such curling, while C and D represent corners where curling has been restrained by anchoring (see text).*

Concrete slabs constructed on masonry walls will behave the same way. In Figure 1, A and B represent two such corners, while C and D represent corners that have been anchored as discussed later. Normally all four corners would curl upward and undergo the cracking indicated in Figure 2. Park and Gamble (Reference 1) say that if slab corners are "not positively held down . . . [they] will lift off the support." Ferguson (Reference 2) recommends that "slabs should be anchored down (to a substantial mass of masonry . . .) and have reinforcing steel provided for the restraining moment thus

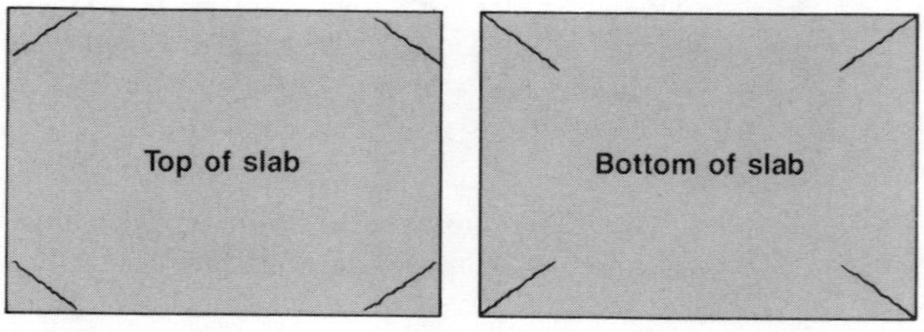

Figure 2. *Corner crack pattern observed in simply supported slabs.*

developed." If the slab corner is attached incorrectly,. however, it will cause horizontal cracks to develop in the wall.

There are two possible solutions. The first is to anchor the slab corners into the masonry to a depth sufficient to prevent the slab from lifting the top of the wall and cracking it. Corner D of Figure 1 represents such anchorage. Corner C represents a slab that caused a horizontal crack at point E near the top of the wall because it was not anchored deeply enough. The other solution is to wait for the slab corners to deflect upward. Then either tuckpoint the joint or, if the slab loading can vary, provide a movable joint between the wall and the slab.

Either solution can generate problems, so be careful. No matter which solution is used, special reinforcement is required in the corners of the concrete slab to prevent the kind of slab cracking shown in Figure 2.

References

1. Park, R., and Gamble, W. L., *Reinforced Concrete Slabs,* John Wiley & Sons, New York, New York, 1980, pages 30-31.

2. Ferguson, P. M., *Reinforced Concrete Fundamentals,* Fourth edition, John Wiley & Sons, New York, New York, 1979, page 353.

Curling of slab over vapor barrier

Q. *On a recent project, the corners of a 5-inch concrete slab on grade curled upward 3/8 inch. Tapping with a hammer on the corners produced a hollow sound. The slab, which contained welded wire fabric 2 inches below the top surface, had been placed on a 4-inch layer of crushed stone over a vapor barrier. Why did the slab curl? Some engineers tell us to place the vapor barrier directly under the slab and others require stone or sand between the barrier and slab. Which is correct?*

A. A slab on grade deflects upward because of a variation in moisture content throughout the thickness. This is the result of more rapid drying from the top than the bottom. The drier top therefore shrinks faster than the bottom, causing curling.

When there is a vapor barrier directly under the concrete it creates a greater difference in moisture content and a greater difference shrinkage, and hence more curling, than otherwise. If a vapor barrier must be used it is considered better practice to place a gravel layer between the slab and vapor barrier. Since moisture moves from the slab in both directions, curling is minimized.

Curling can still occur even if a gravel layer is placed. On one job constructed with a gravel layer, the corners were reported to have deflected upward 1/2 inch. But there were two problems. First, cold weather required the slab to be in an enclosed area which had to be heated to 75° F for three days. The heaters accelerated the rate of evaporation from the top. Second, the gravel had been saturated by snow before it was placed above the vapor barrier. The combination of drying from the top and saturation below the slab caused an extreme difference in moisture content from top to bottom, a prime condition for curling.

Anything that will help reduce the variation in slab moisture content will reduce curling. ACI 302.1R, "Guide for Concrete Floor and Slab Construction," provides a complete listing of measures that can be taken to minimize curling. This useful publication is available from the American Concrete Institute, Box 19150, Detroit, Michigan 48219.

Discoloration

Treatment of efflorescent floor

Q. *We have lived in the tropics in Central and Latin America where concrete floors are common and all I've ever seen done to maintain them is to swab and mop the floors with diesel fuel. However, we now have a 25-year-old-house in the Bahamas in which the 3/4-inch-thick concrete tiles set on the concrete slab have a hard, white calcareous deposit on them, especially under appliances, rugs, and furniture. The ground under the slab is mostly limestone and the floor probably has no moisture barrier under it. The tiles are scratched from many years of abuse and no maintenance. The tiles themselves are beautiful, however, and I would like to restore and maintain them.*

I was advised to acid wash a test area, clean with proprietary cleaners and degreasers, then apply a urethane sealer and finally liquid wax. The acid wash discolored the test area unevenly, so I removed the deposits with dry abrasive scrubbers, then applied the prescribed materials. After two weeks the whole surface discolored from urethane film, so I removed everything again with the abrasive scrubbers. The white deposits started building up again during the rainy season.

Can you tell me how to remove the white deposits and how to seal the tiles with an oil or breathable sealer?

A. It is possible that moisture is continually moving up from the soil, through the slab and tiles, bearing dissolved materials and leaving them on the surface when the water evaporates. You could determine if this is happening by making the following test during a dry period of the year when the relative humidity is low enough so that atmospheric moisture does not condense on the floor. Remove a tile and whatever mortar or adhesive is sticking to the concrete with a black, nonfading felt-tip marking pen to provide good color contrast. Observe the marked area periodically to see if a white deposit is building up on it. If so, moisture is moving up from below and bringing with it salts from the soil or slab or both. (In some locations near the ocean, seawater has been used in the original concrete mix proportioning, thus loading the concrete with sea salts.) You may have to find some method of stopping this continual upward flow of moisture with its virtually inexhaustible supply of salt.

On the other hand, the moisture movement may be mainly or exclusively the result of the moisture that condenses on the floor when the humidity is high and is then absorbed by the tiles. This moisture dissolves some salts from within the floor and then when the humidity goes down the moisture moves up the surface again. There it evaporates and leaves its dissoloved materials behind as the white deposit. Any calcium hydroxide and calcium bicarbonate present in the deposit react with the atmospheric carbon dioxide to form calcium carbonate. This won't be redissolved and carried back down by subsequent condensation because it is practically insoluble in water. In this way an increasingly thick white deposit builds up.

If this is what is happening, the main effort might be to clean the top surface thoroughly during the dry season, preferably with a scrubbing machine that doesn't require abrasive. Then you could treat the tile surface with a material capable of preventing the absorption of the moisture that condenses during humid weather. A good wax might work, or possibly even a treatment with a generous amount of the diesel fuel or mineral oil that appeared successful in the other tropical areas where you've lived. If the moisture doesn't go in, it won't be bringing back its load of salt.

Efflorisence problems

Q. *How can you avoid problems with efflorescence on concrete surfaces?*

A. Efflorescence occurs when salts in the concrete or the immediate environment dissolve in water which then carries them to the surface and leaves them as a whitish deposit when the water evaporates. The salts may have been present in mixing water, for instance if seawater was used in making the concrete. To control that problem, avoid using mixing water that contains excessive salts. Byproducts of cement hydration, such as calcium hydroxide, can also cause efflorescence. This problem can be solved by using fly ash as an admixture in the concrete. The fly ash reacts with calcium hydroxide to form a compound that is insoluble in water. Deicers applied to bridge or parking decks and salts carried by water coming into the concrete from adjoining soil are other sources of efflorescence problems. Adequate curing, use of a concrete with a low water-cement ratio and good waterproofing practices where needed make concrete more watertight and efflorescence less likely.

Color differences in formed concrete surfaces

Q. *Discoloration mars the appearance of exterior concrete on one of our institutional buildings here in the Southeast (see photo). What causes such poor looking concrete? Can anything be done to even out the color?*

A. Without knowing more about construction practices used, we can only speculate about the causes. Some of the discoloration appears to be related to formwork. The dark horizontal lines are caused by water leakage at form sheathing joints. Leakage draws cement and other fine particles toward the joint causing a dark discoloration.

Some of the other discoloration may have been caused by differences in form sheathing absorbency. Absorbent forms darken the surface concrete because they absorb water, reducing the water-cement ratio. Variable absorbency of unsealed plywood, for instance, can cause blotchy looking surfaces. Differences in type of form release agent, or in application rate, may also account for the color variations.

Some of the color differences may have originated at the batch plant. Color differences are likely if concretes made with cements from different sources are used in the same structure. Admixtures such as calcium chloride or silica fume may change the concrete color, as will some fly ashes.

Weathering stains also may have caused some of the discoloration, and in your area mildew may be the culprit. Photos of the building immediately after it was built would help to establish if the mottled appearance was caused by weathering or mildew. Stain removal methods may correct some of the discoloration. Pressure washing with a mixture of water, bleach, and detergent has been used for mildew. Other stain removal techniques are described in *Removing Stains from Concrete*, a collection of articles reprinted from *Concrete Construction*. It can be ordered from The Aberdeen Group, 426 South Westgate, Addison, Illinois 60101.

Discoloration caused by construction practices can't usually be corrected by stain removal methods. Painting or coating is an alternative method, but you will still have to clean the surface first.

Bluish stains will fade

Q. *On an architectural concrete job where we used plastic-coated plywood to get an especially smooth finish, we were shocked to find random bluish stains marring what should have been a beautiful, smooth surface. On areas near form joints and where plain plywood was used, the concrete color was mostly normal. What could have gone wrong?*

A. From time to time bluish-green or greenish-blue discoloration is reported. Usually it is linked to iron sulfides and ferrous oxides found in blast furnace slag used either as aggregate or in the manufacture of slag cement.

This question is quite similar to one reported nearly 20 years ago (*Concrete Construction,* March 1968, page 80), even to the irregular, free-form shapes of the discolorations. The stains seem to be more severe where the forms are more airtight.

The unwanted color generally will fade with age where the concrete is exposed to indoor drying or outdoor weathering conditions. If it is necessary to speed up the decolorizing action, wash the surface with a 3 percent solution of hydrogen peroxide as early as possible after removing the forms.

Minimizing concrete color variations

Q. *What causes color differences in a cast-in-place concrete wall?*

A. Cement is usually the main contributor to the color of concrete. Cements from different plants almost always show color differences. Increasing the cement content or decreasing the water-cement ratio usually darkens the color. Water absorption into the form face while the concrete is hardening may also cause darkening. Concrete that has been allowed to dry quickly after hardening is lighter in color initially than concrete that has been kept moist. As a result, removing formwork at significantly different ages after placement can produce color differences in a wall. However, the differences become less marked with time.

To minimize color variations, purchase all cement for the job from a single plant. Don't vary mix proportions, especially cement and water content. Seal form faces to prevent absorption and tape joints to prevent leakage. Then strip all of the forms at about the same time after placement. The American Concrete Institute's *Guide to the Selection and Use of Hydraulic Cements* (ACI 225R-85) lists several other factors that may cause variations in concrete color.

Concrete turns white

Q. *What causes our concrete to turn white in color?*

A. Some cements produce a lighter colored concrete than others because of differences in chemical composition. If you have changed cement suppliers the color of your concrete could also change. Concrete color is also related to the water-cement ratio and how well the concrete is cured. The higher the water-cement ratio or the shorter the moist curing period, the more likely it is that your concrete will end up lighter in color.

Discoloration of white concrete by curing

Q. *Job specifications require us to water-cure columns. However, the columns are made with white cement, and water-curing is causing discoloration and staining. It seems to me that I have read that curing at 80 to 85 percent relative humidity is satisfactory for columns. Wouldn't it be satisfactory for us not to cure the columns at all, since the weather has been fairly damp and rainy?*

A. The following comment comes from Section 2.8 of ACI 308.81, Standard Practice for Curing Concrete. "During rainy or foggy weather little or no effort is needed to achieve curing, although protection of the surface against washout or erosion in heavy rainfall may be needed. For a very low humidity environment particular care should be taken to prevent excessive moisture loss from the concrete." Section 2.10 says "Natural curing from rain, mist, high humidity, low temperature, moist backfill, etc., may be regarded as sufficient to provide ample curing when its effect is at least the equivalent of keeping the concrete moist for the first 14 days if made with Type II cement, 7 days if made with Type I cement, or 3 days with Type III cement, if kept above 50° F (10° C), unless otherwise prescribed in the project specifications." Natural curing would seem to be practical on your project, so long as the moist weather continues, provided the engineer who wrote the job specifications agrees. Otherwise, you may want to use some kind of curing enclosure that does not bring the curing membrane into contact with the concrete surface, which could also cause discoloration.

Color loss from sandblasting

Q. *In a mockup of an architectural concrete panel, the color of the aggregate seemed to be lightened more than we expected when we gave it a light sandblasting. Is there a way to diminish this?*

A. Try a high pressure water jet with adjustable pressure control operating at 8000 psi or more, using 10 percent sand in the water. Add the sand to the water tank and keep it stirred so it will stay in suspension.

Light etch retarder to avoid color change

In the Problem Clinic item on page 440 of the April 1984 issue, it was recommended that water blasting be used with a small amount of sand in the water as an alternative to sandblasting if sandblasting is causing a change in color. A reader sends another suggestion:

The use of a light etch retarder instead of sandblasting will not change the color, texture, or configuration of the aggregate at all. In addition, there are ecological and economical reasons for this choice.

Black surface discoloration

Q. *Normal mixing and placing procedures were followed on some recent jobs but the concrete had a tendency to blacken on the surface. Could any variations in city water treatment have caused the problem?*

A. City water treatment is a most unlikely cause. A very common cause is known as "trowel burnishing," which sometimes occurs when flatwork is darkened (usually unevenly) by late, vigorous troweling. It seems to be caused partly by abrading metal from the trowel and partly by a reduction in water-cement ratio at the surface.

When the dark discoloration is uniform, it is perhaps caused by a low water-cement ratio in the mix.

Discoloration from forms

Q. *The exposed smooth walls of a commercial office building are not uniform in color. The plywood forms used were new when the job began. We noticed that the forms absorbed a considerable amount of form release agent the first time it was applied to them. As time went on, less form release agent seemed to be needed. The walls also seemed to be a little bit more uniform in appearance towards the end of the job. What could have caused the nonuniformity of color?*

A. Your description suggests that you used too highly absorptive plywood. Nonabsorptive plywood should have been used. Absorptive plywood removes moisture from the concrete mix at the surface next to the form. Some calcium hydroxide from hydration of the cement dissolves in the mix water and this also becomes absorbed. Loss of this light-colored portion of the mix causes the concrete to become darker. The solution to this problem is to use a less absorptive plywood.

White discoloration predicted

Q. *We are bidding on some city sidewalks where the specifications require that plastic sheet be used for curing. We know that this ordinarily causes white discoloration and we don't want to be held responsible. Is there any way to prevent the discoloration?*

A. You can't prevent white discoloration without going to another curing system—either a liquid membrane curing compound or some kind of continuous wet curing. We suggest that you show the city a copy of the four-page article "Discolored Concrete Flatwork," published in our November 1967 issue, page 417, and argue that the city must either accept white discoloration, at least initially (white discoloration does slowly diminish over a considerable period of time), or approve a different curing method.

Tan discoloration of white stucco

Q. *A stucco was intended to be white but is light tan in color. The stucco mix was made with one bag of white portland cement, one bag of white masonry cement and three 5-gallon buckets of masonry grade sand mixed with torpedo sand. The white portland cement and white masonry cement have been checked for color and have been found to satisfy the color requirement. Subsequent efforts to reproduce the same tan color have failed. What could have caused the tan discoloration?*

A. It sounds very much as though the mix became misproportioned to the extent that the sand determined the color. There also might have been excess fines in the sand. We suggest that you determine the cement content of the stucco by chemical or petrographic analysis to see if it conformed to the mix design. If it did not conform, the sand content may have been high or may have contained an excess of fines.

Midstream horse switching

Q. *We are the architects on a concrete house that is to have an exposed-aggregate finish produced by sandblasting. A 12-foot-high continuous wall, which comes out of a hill, is more than 200 feet long and is being placed in sections, using 2500-psi concrete. Now, when an infill section is about to be placed, the owner is panicking. He is worried that the strength is not sufficient for so high a wall and wants to switch to 3000-psi concrete. Our design shows the higher strength is not needed and the mix has been placing well up to now. We are concerned that additional cement would cause an objectionable color difference. This wall is the most obvious part of the house.*

A. You would be inviting trouble with appearance if you were to change the mix at this point. If you wall is already satisfactory in all respects it would be best to stick with it.

Change in cement changes flatwork color

Q. *Construction contingencies caused us to end up with a one-panel gap in a driveway. Later we filled this in with another load. The photo shows the result. We have learned that*

the portland cement in the last load came from a different cement plant from the rest. Could a change in cement cause this much change in color? Is there anything we can do to correct it?

A. A change in brand or type of cement almost always causes a change in color, not always as dramatic as in your experience. Making the exceptional panel match the others would be difficult. You might try a stain, in consultation with the manufacturer, but success would probably require some artistry to get a good match, and it is not certain that the color would continue to match after weathering.

If it is important enough to have a precise color match it could be accomplished by taking out the whole panel and replacing it with concrete made with the same cement and mix proportions as the adjacent sections, and curing it in the same way. Any changes in materials might lead to another difference in color, though not likely as marked as before.

A less satisfactory solution would be to make a grout mix using 1 part of the correct cement and about 3 parts by volume of minus 30-mesh sand and water. This could be scrubbed into the predampened surface and a curing compound applied. (If there is still any curing compound on the original surface, a couple of months should be allowed to elapse before coating so that the curing compound will disappear. Any dirt that accumulates in the meantime should be scrubbed off and the surface thoroughly rinsed.) A trial mix would have to be made, applied to a limited area, cured 3 days and allowed to dry. If too dark, the mix could be adjusted by replacing some grey cement with white cement. If too light, the cement content could be increased slightly. The problem with this treatment is that the coating might wear off before long.

Another possibility is to accentuate the difference with plantings or decorative side strips to make the effect look intentional.

Scaling and Spalling

Spalls caused by powder-actuated fasteners

Q. *What causes concrete to sometimes spall when powder-actuated fasteners are used? Is there a solution for the problem?*

A. There are several possible reasons for spalls

- The fastener may deflect and cause a spall if it hits a rebar or a coarse aggregate particle.
- The concrete may be too strong. If the strength exceeds about 7000 psi, concrete may be too brittle to permit the plastic deformation that allows the fastener to penetrate.
- The powder charge may be too large.

Try using a smaller powder charge. Start with the smallest charge and work up to the power level needed for embedment. A spall guard accessory is available for most guns. Attached to the nozzle, it exerts a confining pressure on the top concrete surface to minimize spalling. As a last resort try drilling a small pilot hole before shooting the fastener.

Keeping concrete particles out of food

Q. *Our company has a room in which machines are located directly below the hoppers of some 15-foot-square reinforced concrete bins that hold raw popcorn. Sometime in the past the bins were sprayed on the inside with an epoxy coating. Now small rock-size chunks of concrete are breaking out of the walls of the bins and falling with the popcorn into the machines. The concrete in the walls is very hard, but the exterior walls of the building show rust over the rebars. We think vibration from the machines may be causing the particles to break out. What can we do to stop the concrete particles from getting into the popcorn?*

A. You need to find out for sure what is causing the particles to break out. If there is a cause of moisture buildup behind the coating you might be getting popouts from traces of chert or other expansive aggregate. Vibration from the machines doesn't seem a likely cause of popouts, though perhaps vibration could loosen concrete particles adjacent to any cracks that exist. A concrete petrographer could identify the cause by examining some particles and a sample of the area from which they came.

The kind of repair needed will depend on the cause. Very possibly you will want to recoat the repaired area by applying an epoxy or polyurethane coating. To find the names of manufacturers of such coatings suitable for use on concrete look under "Coatings, Wear Resistant" in the Buyers' Guide in our most recent December issue.

Spalling at keyway

Q. *A warehouse floor has suffered spalling at joints that contain a metal keyway. What caused the spalling?*

A. Without knowing more it is impossible to say, but spalling does occur frequently at keyed joints that are subjected to hard-wheel traffic. Although this can happen to any keyed joint it seems to occur more frequently in joints formed by metal keyways that are left in place (that is, in control joints, since keyways are removed from construction joints before concreting the adjoining slab). There can be any of several causes:

- Keyways of any kind cause a small vertical gap between the male and female portions when the joint opens. The pounding in this gap from repeated impact of traffic can cause spalling.
- The slab may be too thin. The small amount of concrete above the keyway in a 4- or 5-inch-thick slab offers less resistance to spalling than in a thicker slab. If keyways are to be used in such circumstances it might be well to thicken the edge to 6 inches and use a keyway made for a 6-inch-thick slab.
- The trouble can be caused by surface subsidence during bleeding of a mix that is too wet. The finished surface of most of the slab is lower than the top of the metal keyway. Even though the finisher trowels the nearby surface up to the level of the top of the keyway the joint presents a high ridge to the impact of oncoming traffic.

Spalling and keyways

One reader, John E. Hutchinson from the Form-A-Key Products Division of Cardinal Manufacturing Company Inc., had some additional advice on spalling and keyways.

In the July issue, page 621 a reader raised a question about spalling at a keyway. Your answer stated that spalling occurs frequently at keyed joints that are subject to hard-wheel traffic and that it occurs more frequently in joints formed by metal keyways that are left in place.

We find that the major cause of spalling at the joint can also be from using a mix that is improper for the prevailing construction conditions and the finishing methods employed. There are on the market today, however, several different designs of the metal joint—some with better design advantages than others.

The metal joint could have either alleviated or contributed to the problem, depending on which brand was installed.

I hope this information will expand your knowledge of some of the concrete accessories that are available today.

Cause of spalling or scaling

Q. *Our company placed about 300 cubic yards of concrete in curbs, sidewalks and aprons in a business district whose roads are serviced by the state. The concrete was certified to be 3500 psi and air-entrained. Welded wire fabric, size 6 x 6—W1.4 x W1.4, was used. A cure-seal spray was applied at completion and the concrete was also covered with polyethylene sheet for one day.*

Concrete was only 4 months old when spalling occurred in various sections totaling about 20 cubic yards. Salt from cars crossing over 6-inch aprons at the parking lot entrance caused spalling of new concrete. Could this problem have been prevented, and if so, how? Who should be responsible for this problem?

A. Possibly the problem is scaling. This is similar to spalling except that the concrete tends to come off in thin layers. Scaling is often the result of the action of deicers during periods of freezing and thawing.

The responsibility could lie with the specifier, the concrete supplier or the contractor, depending on what caused the trouble. Based on what you've reported it does not seem as though the concrete should have spalled or scaled, so there must be some condition that is either unknown or incompletely considered.

You said that the concrete was certified to be air-entrained, which we take to mean that the air content was not actually tested at the job site. You didn't indicate whether a definite air content had been specified. For a severe climate like yours, the average air content specified should have been 5.5 percent if the maximum size of the coarse aggregate was 1½ inches, 6.0 percent if the maximum size was 1 inch or ¾ inch, 7.0 percent if it was ½ inch, and 7.5 percent if it was ⅜ inch. If the concrete contained less air than required for the maximum aggregate size, it could have contributed to scaling or spalling.

Sometimes the job concrete contains less air than specified because either the air-entraining agent was accidentally left out while batching or the air was partially lost at some later time. It's possible for some of the initial air content to be lost during transport and delivery if the concrete contains fly ash, particularly if the fly ash has a high carbon content. Other possible causes of loss of entrained air are adding water to the concrete in the mixer to increase the slump, finishing the concrete while there is bleed water on the surface, or spraying water onto the surface of the concrete during finishing. One of these troubles could be the cause of the spalling or scaling. Sometimes one or more of these causes is unknown to the contractor and does not show up in the job records.

Although good curing requires retaining moisture in the concrete, there is a potential difficulty if the concrete is placed late in the fall. If the curing keeps the moisture content of the concrete high during the winter season, it may contribute to scaling after salt is applied. The recommendation is often given that liquid membrane curing compounds should not be applied to concrete placed late in the fall, since these compounds may retain a high moisture content for an undesirably long period during the freezing and thawing season, thereby contributing to scaling if the concrete is low in air content or high in water content. For this reason it is also suggested that concrete be allowed to cure and then dry at least 28 days before any salt is applied. Others recommend that applications of deicer salt be prohibited during the first winter.

By testing a hardened sample taken from the pavement a petrographer could ascertain the air content of the concrete and also determine whether the water-cement ratio of the cement paste in the surface layer is higher than in the rest of the concrete.

Scaling of exposed aggregate sidewalks

Q. *Some exposed pea gravel aggregate sidewalks were placed in the fall near the shore in the state of Washington, using a 6-bag mix with hot water and an accelerator. The finish was beautiful, and there was no damage from some snow and below-freezing temperatures the first winter. In the following summer an aggregate glaze was applied. Several months later, in November, we had 2 feet of snow and below-freezing temperatures. Rock salt deicer was applied to about two-thirds of the sidewalks. Soon afterward we observed areas about a foot in diameter that had spalled or scaled. In areas where no salt had been applied there were pea gravel popouts. Tests of cores taken from areas subjected to salt showed strengths of 2416 and 2860 psi, and a core from a nonsalted area 4425 psi. A testing laboratory reported that the strengths were above the 2000 psi required by the Uniform Building Code, attributed the scaling to the application of deicer salts, attributed the popouts to the exposure of the aggregate and to water absorption by the aggregate, and recommended that in future work the concrete be air-entrained and the surface sealed. We would like to know if you agree.*

A. Based on the information you give, the Uniform Building Code requirements appear to be what govern strength on this job. The strength tests of cores show that the concrete met that part of the specifications. As far as resistance to freezing and thawing and to salt scaling is concerned, however, an adequate air content is even more important than strength. In addition to its strength requirements, the Uniform Building Code does require that concrete contain entrained air if it will be subject to freezing temperatures when it is wet. It sets forth the amount of air required on the basis of the maximum size of the coarse aggregate. Apparently your concrete was not air-entrained, and this could have been a major cause of the trouble.

Without knowing the nature of the aggregate glaze and its performance record, it's impossible to say whether it helped or hindered the performance of the concrete. A sidewalk retains some of its original mixing water for a long time. Subsequently, additional water moving up from the soil below is absorbed by the concrete. If the glaze is made of a material that has low permeability to water vapor, it may not allow water to evaporate from the concrete and under some circumstances this can make the concrete, if not air-entrained, more vulnerable to freeze-thaw damage. Application of deicers to non-air-entrained concete causes scaling, and usually more rapid deterioration, than freezing and thawing alone.

Popouts are usually caused by certain highly absorptive particles that may be present in small amounts among other kinds of aggregate particles. The American Concrete Institute publication *Cement and Concrete Terminology* defines the word popout as meaning "the breaking away of small portions of a concrete surface due to internal pressure which leaves a shallow, typically conical, depression." At the bottom of the depression there is usually a piece of porous aggregate that has expanded with moisture and pushed out a cone-shaped piece of mortar. If your concrete contained some highly absorptive particles, they may be part of your problem. If it's a true popout you can usually see a piece of broken aggregate at the bottom of the cone-shaped crater, but if there's any question about it a petrographer could tell by examining the aggregate particle itself.

In future installations in your area, where freeze-thaw exposures can be considered moderate, it would be important that you achieve an air content of 6.0 percent for concrete made with 3/8-inch maximum size aggregate like pea gravel. If you want to protect exposed aggregate further by using a coating, it would be important to use one that is known to be relatively permeable to water vapor.

Scaling concrete

Q. *I would like to know how contractors cope with concrete slabs poured in the summer and sealed with a sealer. I've used two different types of sealer but in both cases I got salt scaling the following spring where salt had been thrown around. I know it is hard to cope with salt. I know my problem has not been the concrete because I used a 6-bag mix.*

A. You didn't mention whether or not the concrete was air-entrained. If it didn't contain entrained air, the application of salt combined with the exposure to freezing and thawing would cause the concrete to scale. This would be true regardless of the cement content, sealing, time of pouring, or any degree of care exercised.

To produce air-entrained concrete, an air-entraining agent is introduced into the concrete at the time it is batched. As the concrete is mixed, microscopic air bubbles are formed in the cement paste. They provide space for water to expand within the concrete when it freezes. Without these air bubbles, the expansion of water during freezing causes the surface concrete to loosen and disintegrate.

We are sure you will find that all paving work done by the state or any municipality in your area requires air entrainment. Usually the air content is specified at 4 to 7 percent or 5 to 8 percent. Any ready mixed concrete producer will be familiar with entrained air. When you order concrete for any outdoor flatwork you will want to make certain that the supplier understands that you want air-entrained concrete.

With adequate air content it is still necessary to use good construction practices. An extremely wet mix, finished too early or too much, may greatly reduce the effectiveness of the air entrainment. Concrete placed late in the year, not cured nor subsequently allowed to dry out, may still be vulnerable to some scaling. A fair guideline is to be sure that the concrete is not exposed to salt and freezing and thawing for at least 30 to 60 days after finishing.

The value of sealers is questionable for concrete properly constructed, cured and dried. Boiled linseed oil has been used to protect pavement placed late in the year with some good results. Many other sealers may have a reverse effect. They serve well as curing compounds but may then serve badly by preventing the concrete from drying out after it has cured. When using boiled linseed oil it is necessary to take the precaution of allowing the concrete to dry out after curing, before applying the linseed oil.

If you use the 6-bag mix that you have been using but with the right amount of entrained air, and the construction practices that are outlined here, you really should be able to cope with the scaling problem.

Use of jitterbug and scaling

Q. *We have a sidewalk that has scaled. The contractor who built it didn't seem to be competent; he didn't even use a jitterbug. Doesn't this warrant our demanding that he repair or replace it?*

A. Surface scaling can result from any of a number of conditions or practices, including an inadequate air content, excessive water in the mix, finishing while bleed water is still on the surface, adding water while finishing, inadequate curing, or applying deicers before the slab has dried adequately after curing. Actually, using the jitterbug is not recommended unless the slump of the concrete is less than 3 inches or the concrete is made of lightweight aggregate. Otherwise using the jitterbug could *contribute* to scaling, not prevent it.

Establishing responsibility for scaled sidewalk

Q. *We put in a sidewalk for the city in November. By the following summer it had scaled. The city ordered tests made of the air content and found it to be low. Then it accused us of using non-air-entrained concrete. The concrete was air-entrained and it was sealed with a cure-and-seal agent. However, the city salted the sidewalk heavily during the winter in spite of our warnings to them not to do it. Is there a reliable way to measure air content on hardened concrete?*

A. ASTM C 457, "Standard Practice for Microscopical Determination of Air-Void Content and Parameters of the Air-Void System in Hardened Concrete," gives reliable results.

In case the air content turns out to be low, you should know that there are a number of possible causes for the concrete in a sidewalk having less air than specified. The concrete may have been improperly batched or mixed; or it may have lost its original air content by improper handling, by overwetting or by working too early on the job. The surface of flatwork is particularly susceptible to loss of air by overwetting or by working too early.

Be that as it may, any flatwork should be given at least 28 days to dry (after curing) before any deicer is applied. ACI Committee 302 recommends 3 months drying after curing. The Portland Cement Association has recommended that liquid membrane curing agents not be used on flatwork placed in the late fall. They suggest that the flatwork be cured instead with a waterproof covering such as polyethylene.

Don't seal new sidewalks

One reader, Raymond J. Schutz, Technical Director of Protex Industries Inc., wrote to explain the probable cause of scaling in the previous Problem Clinic item.

A Problem Clinic item on page 714 of the September issue discussed establishing responsibility for a scaled sidewalk. This note is not in regard to establishing responsibility for the scaling but to point to the probable cause.

I note that the concrete was sealed with a so-called cure-and-seal. Such materials can do an excellent job of curing and sealing, but if the product does a good job of sealing it can cause premature scaling of concrete placed outdoors on grade in a freezing climate.

If the surface of the concrete is sealed, there is no escape for moisture. The concrete under the seal can become critically saturated and when the trapped water freezes, the concrete can scale (sometimes on the first freeze) even if the concrete is air entrained. One can see this phenomenon by lifting up a rubber mat placed on a sidewalk. It generally will be wet underneath.

Concrete on grade in a freezing climate should not be sealed. A linseed oil impregnation (which breathes) would have been a better solution, or an ordinary curing compound meeting ASTM C 309 would also have been a better choice.

Cure, then age before salting

Q. *What's the scoop on preventing salt scaling after deicers have been applied during the first winter on slabs built in the late fall?*

A. This subject has been discussed in the Portland Cement Association publication, *Concrete Technology Today*, Number 3, September 1980. A slightly condensed version of that discussion follows.

Avoid the use of deicers the first winter if possible and apply cinders or other abrasive materials instead to eliminate slipperiness. If this is not practical, adopt a schedule of late fall paving designed to minimize the deteriorating effects of using deicers. Before deicers are applied, the strength level specified for the concrete should be attained and some additional period of air drying will be necessary.

In planning late fall work, one should consider the effects of curing, drying, age and air content on the durability of strength of the concrete.

Curing. In the PCA laboratories, surface scaling tests were made by Klieger to provide a basis for deciding on minimum curing periods before deicers are applied. Both non-air-entrained and air-entrained concretes were mixed at 40° F to simulate cold weather construction. The test specimens were continuously moist cured at 70° F and lower temperatures for different periods of time ranging from 1 to 60 days.

None of the non-air-entrained concretes showed a satisfactory degree of resistance to surface scaling, regardless of the curing temperature and the duration of the curing period.

The tests indicated that for air-entrained concretes to develop resistance to surface scaling a minimum curing period is required (see table).

The table shows that the minimum curing period for scaling resistance is not alike for all concretes. It depends on the type of cement and curing temperature and whether accelerator was used. The minimum curing period at any temperature is approximately the length of time normally required to develop a compressive strength approaching 4000 psi.

Air drying. In another study on the effect of curing and subsequent air drying, Verbeck and Klieger concluded: "A period of air drying of the concretes prior to the start of scaling tests increased the resistance to surface scaling."

Jobsite concrete rarely obtains curing comparable to continuous moist curing in the laboratory. Generally, after a brief curing period—or none at all—the concrete is exposed to drying conditions.

These laboratory tests showed that in almost all instances a 14-day period of air drying following moist curing increased the resistance to surface scaling of air-entrained concretes. For non-air-entrained concretes, air drying improved durability somewhat.

However, the scaling resistance of the surfaces was not satisfactory. The air-dried air-entrained concretes showed very good resistance to scaling through 200 cycles of severe freeze-thaw testing—an indication that these concretes would be durable and have an excellent service record in the field.

MINIMUM CURING PERIOD FOR ADEQUATE SCALING RESISTANCE
6-bag mix, 5- to 6-percent air content

		Made and cured at 70 degrees F		Made and cured at 40 degrees F	
Cement type	Calcium chloride, percent	Minimum cure time needed, days	Strength developed, psi	Minimum cure time needed, days	Strength developed, psi
I	0	7	3800	15	3900
	2	7	4100	7	3700
II	0	7	3250	12	3400
	2	7	4500	7	3300
III	0	7	3700	7	3750

Liquid membrane-forming curing compounds will delay slab drying. In late fall work, this is not an advantage and curing with a polyethylene sheet may be more beneficial because the sheet is entirely removed for the subsequent air drying.

Age. Hansen (Universal Atlas Cement Company) made tests to determine how age affects the resistance to frost and salt scaling the first time deicers are applied. Slab specimens were placed in the field at intervals from August 8 through December 4. The concretes had air contents of zero, low, and normal. After 7 days moist curing, the slabs were air dried until December 3, when they were covered with water to a depth of $^3/_8$ inch. Slabs placed December 4 were covered with water December 11 and had no opportunity to air-dry before being subjected to severe freezing. On through the winter the slabs were kept covered with water and whenever it was frozen solid the ice was melted with calcium chloride deicer. A total of 55 applications of deicer were made.

Except for the slabs that were 3 and 4 months old at the first freeze, those made without entrained air were completely scaled in less than 15 cycles of freezing and thawing. The concretes with low air contents were completely scaled in less than 55 cycles of freezing and thawing.

Hansen concluded that all of the normal air-entrained slabs that were at least 24 days old (7 days moist curing plus 17 days air drying) before being subjected to cycles of freezing and deicer thawing probably would give satisfactory future service. Each of Hansen's scaling-resistant slabs had developed a compressive strength of over 4000 psi by the first freeze.

Though laboratory test conditions do not duplicate those of nature they can be the means of developing helpful information. A person deciding whether or not it will be safe to use deicers on new concrete slabs, or to permit salt brine to drip from vehicles parked on the surface, may want to base the decision on:

- frequent testing of concrete for adequate air content
- whether the concrete has been cured long enough to achieve 4000 psi compressive strength
- whether there has been a 14-day period of air drying (or preferably longer) following moist curing
- whether the slab is at least 28 days old (older is better) before deicers are first used. This is equivalent to 7 days moist curing for strength gain and 21 days of air drying.

Further discussion can be found in Portland Cement Association *Research Department Bulletins RX82 and RX83, and in W. C. Hansen's article in the ACI Journal*, January 1954, p. 341.

Deicers on parking garage decks

Q. *Do you advocate that parking garage decks be designed to protect against deicer attack even if they are enclosed? Is there enough dripping of salt water off of parked cars to cause trouble?*

A. Where salts are used as deicers on streets, large amounts of salt will be brought into parking garages. Even if no salt is purposely applied to the deck for deicing, a considerable amount of salt will accumulate in the concrete over a period of time unless something effective is done to protect the concrete against absorbing it.

Superficial marks and durability

Q. *A contractor applied a broom finish to a driveway but because the weather was cold he placed kraft paper over it and covered it with salt hay. In some places the paper became pressed into the surface enough to make the surface smooth. Will these smooth areas be more subject to scaling and spalling?*

A. Air entrainment was not mentioned. If the concrete mix had a sufficient air content and if the concrete was placed at a slump of no more than 4 inches the more critical considerations for freeze-thaw durability have been taken care of. (It would also help if the concrete had a chance to dry for 28 days between the end of curing and the beginning of winter, and if no deicers were applied the first winter.) The smoothness of the surface should have virtually no effect on scaling and spalling.

Surface defects

Shallow surface holes that aren't popouts

Q. *We've got some holes that look like popouts in our concrete sidewalk. But you can normally see an aggregate particle at the bottom of most popouts, and there's no particle in most of these. Can materials other than aggregates cause popouts?*

A. Clay lumps near the surface can cause popouts. So can lumps of undissolved calcium chloride, sometimes found in concrete placed during cold weather. These materials might wash out of the hole after the piece of concrete popped out. However, there's another possibility.

Are the holes all the same size? And is there any pattern? Our contributing editor, Bruce Suprenant, found an interesting case of surface damage to sidewalks in Florida (see photo). There were a number of holes in the sidewalk but they were spaced in a regular pattern and were all near the edge. Bruce thinks it's most likely that the damage was caused by a soil aerator (see photo) being pulled over a part of the sidewalk.

Holes in a concrete sidewalk (left) appeared at first to be popouts. However, they were regularly spaced and it's more likely that they were caused by a soil aerator, perhaps similar to the one shown at the right.

Shale popouts

Q. *Last fall we placed a shop floor in an enclosed but unheated farm building here in North Dakota, using heaters during construction and the remainder of the first 24 hours because the outdoor temperatures were in the 40s and 50s Fahrenheit. The concrete was a 5 1/2-bag cement-and-fly ash mix with a 4- to 5-inch slump (estimated, not measured), air-entrained, but with no other admixtures. Within 7 to 10 days we began having a few popouts and in the next 2 or 3 weeks we had many. We attribute the trouble to shale particles. Our state limits the shale particles to 0.7 percent of the aggregate in concrete used for highway work. Is there a way to determine whether the aggregate in our concrete exceeded 0.7 percent shale particles?*

A. Assuming that there is some way in which the state specifications govern your job, there is a way to determine the shale content, though it may be considered a little expensive. It involves taking at least one 4- or 6-inch core containing a popout, plus some chips from around some of the other popouts, and sending them to a qualified petrographer. It would also be desirable to diagram a typical portion of the floor, showing the locations of popouts. For about $500 to $700 the petrographer could determine the percentage of offending shale in the floor.

It should also be said, however, that even if the shale content were below the state requirement of 0.7 percent there could still be an objectionable number of popouts in concrete used for a floor. The 0.7 percent limitation was probably established for highways, where small popouts may not be particularly objectionable.

Type of popout depends on the shale

Bernard Erlin (Erlin, Hime Associates Division, Wiss, Janney, Elstner Associates) has written in response to the item "Preventing Shale Popouts in Floors," which appeared in the October 1985 Problem Clinic, page 879. He points out that our response does not differentiate between two possible kinds of popouts. The question to which we were responding had come from a reader in North Dakota who was having trouble with popouts caused by opaline diatoms in the shale. This problem sometimes occurs in Minnesota, Iowa, the Dakotas and those parts of Canada just north of those states. Bernard Erlin notes:

The first kind of popout is the kind described in the Problem Clinic item referred to. It is caused by the alkali-silica reaction and occurs throughout the world not only with opaline shales but also with alkali-reactive chalcedonic cherts, limestones that contain opal or chalcedonic chert, glassy acidic volcanic rocks such as rhyolitic tuffs, manufactured glass such as bottle glass, and other similar materials. Aggregates containing alkali-reactive silica can be found wherever concrete is used and these materials are not limited by political, geographic or other boundaries.

The second kind of popout that can be caused by shales are those that result from moisture absorption and expansion, and cyclic freezing, even though the shale does not contain opal. Shale is a layered material of variable porosity and permeability. It can absorb water, whether from rain or simply from frequent mopping of a floor. Some shales swell and cause popouts from this uptake of water alone or, when critically saturated, from cyclic freezing. The water-absorption and cyclic freezing mechanisms are different from the one discussed in the October 1985 Problem Clinic item. The alkali-reaction, absorption and cyclic freezing problems are not confined to the several states and the south central portion of Canada mentioned, and can occur whenever the right (or wrong) kind of shale is present and the environmental conditions favor the reaction.

Popouts in new concrete

Q. *In New York State we noted excessive popouts in the flat concrete work which was poured during the latter part of last year. At first we blamed the use of deicing salts. However, closer examination of the damaged portions revealed a piece of either flint or chert aggregate at the bottom of each popout. The popouts occurred only in those portions subjected to deicing salts. Have we made a correct evaluation? Is there a way of counteracting the problem?*

A. Deicing salts do increase the destructive effect of freezing and thawing on concrete that is not air-entrained. It seems reasonable to expect that they also hasten the destruction of porous aggregate particles such as flint or chert. You may not have seen the end of the problem. Popouts normally develop gradually over a period of several years, depending on the exposure, and it is quite possible that eventually you may see additional popouts in areas that have not been subjected to deicing salts.

One method of prevention is to find a source of aggregate that does not contain the offending particles. The easiest way to do this is to check service records of aggregates that are in current use.

Another way, if nonpopping aggregate is scarce, is to place the flatwork in two courses, using the nonpopping aggregate only in the $1^1/_2$-inch or 2-inch top course. Good practices for 2-course construction should be followed, such as those described in ACI 302.1R-80, "Guide for Concrete Floor and Slab Construction."

In a few places aggregate producers have set up beneficiation operations to remove the lighter weight porous particles by heavy media separation. This is a good solution if available.

Coatings to prevent moisture buildup in concrete flatwork have been advocated. These are meant to keep the porous particles from becoming saturated because saturated particles are susceptible to damage from freezing. Coatings that significantly reduce the slip resistance should be avoided and no coating should be used that does not allow the concrete to breathe. Nonbreathing coatings can cause moisture rising from the soil to condense near the concrete surface and cause spalling.

Repairs of popouts can be made by breaking out or drilling out the offending particle and replacing with a patch. Coring or drilling is preferable to breaking because it produces a square shoulder. The patching mix should be relatively dry to minimize subsequent drying shrinkage but the surrounding area should be moistened before patching in order to prevent excessive robbing of moisture from the patching mix.

Some owners are satisfied to live with the problem of popouts and do nothing about them. In some circumstances their appearance is not highly objectionable, and they do not usually diminish the structural quality of the flatwork unless associated with some other problem.

How to fill core holes

Q. *After cores have been taken in a concrete member, what do you use to fill the hole?*

A. Because such a small amount is needed, the patching material is generally hand mixed onsite. Bagged dry-mixed concrete is convenient to use and won't shrink as much as a mortar would. To further reduce shrinkage, make a stiff mix by adding as little water as possible to the dry ingredients. Hand-mixed concrete can't be air-entrained so the patch may deteriorate if it's exposed to freezing and thawing while saturated. A low water-cement ratio makes the patching material more watertight and prolongs its life.

Minimizing bugholes

Q. *We have a continuing problem with bugholes (surface air voids) in concrete walls. We have tried many different vibrating techniques but haven't eliminated the problem. Is this just something that we have to live with?*

A. Committee 309 of the American Concrete Institute (ACI) gives the following advice for reducing the size and number of surface air voids in walls.

- Don't use sticky, oversanded concrete mixes.
- Avoid high viscosity form coatings or coatings that are applied in thick layers. They tend to hold air and water bubbles.
- Reduce the distance between internal vibrator insertions and leave the vibrator in the concrete longer.
- Make a row of vibrator insertions as close to the form as possible, but without touching the form.
- Where appearance of the concrete is very important, hand spading next to the form may be helpful.

Bugholes near the top of a wall

Q. *We've got problems with bugholes in an 8-inch-thick architectural concrete wall that we're building. It's 4 to 6 feet high and the surface looks fine, except in the top 1 foot where lots of bugholes have formed. The air-entrained concrete is colored with an integral coloring agent and is being placed against a plastic form liner to produce a textured effect. We can't vibrate close to the form liner because we're afraid we'll hit it and mar the surface. The concrete is being placed at a 4- to 5-inch slump. We don't know the air content. What can be done to prevent the bugholes?*

A. Try inserting the vibrator at closer intervals and leaving it in the concrete longer. Hand spading concrete near the liner also may help. World of Concrete® seminar speaker Ralph Spannenberg suggests several additional steps that may help. An accumulation of bugholes at the top of a wall indicates that they may be traveling upward as the concrete bleeds. Try using a lower slump concrete to reduce bleeding.

Also check the concrete air content. Most integral coloring agents contain water reducers that also may entrain air. If both coloring and air-entraining admixtures are being added at the plant, bugholes may be caused by too much air in the mix. Adjust the air by cutting the dosage of air-entraining admixture. Try to hit close to the lower specification limit for air content.

Finally, you might try vibrating the top 1 foot of concrete using a form vibrator. After topping out the forms and vibrating the concrete internally, hold the form vibrator against the studs or walers near the top of the wall.

Dusting—what caused it?

Q. *I try to do things right, and I've never had a dusting floor before, but now I do. What caused it?*

A. Check the following possibilities taken from ACI 302.1R-80, "Guide for Concrete Floor and Slab Construction":

- an overly wet mix
- floating or troweling bleed water into the surface
- Clay, dirt or organic materials in the aggregate
- water applied to the surface during the finishing process
- using unvented heaters to protect new concrete from cold weather
- inadequate curing, especially if the weather was dry

Crumbling during edging

Q. *When we use air-entrained concrete we find that the concrete becomes crumbly when we put on our finish edging. What causes this?*

A. This problem is related to the sand gradation and it occurs when the sand is too coarse. This means that there is too small an amount of material that passes the Number 16, Number 50 and Number 100 sieves. If you can get your supplier to provide a sand which contains enough more fine material it should solve your problem. If not, you might improve the concrete by using a mix with a little bit more cement or with a small amount of fly ash in addition to the quantity of cement already in the mix.

Superplasticizer trouble with low-sand mix

Q. *The picture on the next page shows a foamy area that showed up on a 35-foot column after the forms were stripped. On close inspection you can see exposed aggregate, both fine and coarse, in the defective area. In some places you can even see exposed rebar. We had similar troubles on other tall columns. Some 15-foot columns were not quite as bad. The mix contained superplasticizer and the slump after adding this admixture was 9 inches. Is it possible to prevent this trouble when superplasticizer is used?*

A. This kind of trouble is fairly common in mixes that have low sand contents. Low-sand mixes are often used for concrete in large members and they are standard in pavements and dams. Superplasticizer should not be used without making trial mixes that include the superplasticizer with the job materials unless you have prior experience with the same mix. Trial mixes will show whether the proportion of fine to coarse aggregate needs to be increased, and if so, how much. Segregation, poor workability and sand streaking are common troubles when the aggregate proportions have not been properly adjusted for use with the superplasticizer. If you do have trouble, first look at the quantity of the superplasticizer chosen, and then the aggregate proportions. One or both may be at fault.

Repair and Maintenance

7

Concrete

Call in an expert

Q. *The vertical edges of a number of exterior columns of one of our apartment buildings are breaking away, leaving reinforcing steel exposed, and this steel is corroding. There is also some vertical cracking beginning to occur on the surfaces between the cracked edges. What is the best material to use for repairing the spalled areas? Can we repair the cracks by injecting with an epoxy?*

A. It is too early to plan a repair program. The first thing to do is determine what is happening so that you can consider what measures can effectively eliminate the cause. It could be, for example, that the steel in the columns is suffering from galvanic attack intensified by chlorides. If so, you will want to consider what can be done to prevent corrosion from continuing.

This kind of investigation normally requires an expert who can use any of a number of methods to pin down the cause and degree of seriousness of the trouble:

- inspecting and measuring crack widths
- taking cores to determine specific gravity, compressive strength, absorption, percent of voids, and chloride content
- sonic pulse velocity tests of soundness of the interior concrete
- electric potential measurements with a half cell to determine the extent of galvanic activity, if any

Normally the expert would start with the easiest tests and perform only those that are necessary. The choice of repair program would depend on the findings.

Protect finishing operations against windborne particles

Q. *While a slab was being floated and troweled the wind blew polystyrene beads onto the surface. These came from loose fill insulation in an adjacent concrete block wall.*

The finishers could not remove these lightweight particles and the particles got worked into the surface. This marred and pitted the finish. The beads were too numerous to be picked out of the surface by hand. The blemishes promise to be dirt catchers as well as being unsightly. How can these numerous pits be repaired?

A. If it is possible to accept a floor that is slightly thinner and that has an elevation a fraction of an inch lower, you might plane the surface. Otherwise you might repair by power wire brushing the beads out of the surface, then squeegeeing an epoxy over the surface. Some of the methods specified in ACI 503.3, "Standard Specification for Producing a Skid-Resistant Surface on Concrete by the Use of a Multi-Component Epoxy System" may be useful. This appears in the *ACI Journal,* September 1978, page 448.

It would be best to refrain from placing any type of loose insulation that could become windborne on the same day that concrete is being placed. It would also be advisable to make sure that previously placed insulating fill is protected from exposure to wind during finishing operations.

Repairing bead-pitted floors

Guy H. Wilson at Tews Lime & Cement Company had additional information on the Problem Clinic item "Protecting Finishing Operations Against Windborne Particles."

This Problem Clinic item concerned a concrete slab into which beads of polystyrene insulation had been troweled. These had been blown onto the surface by wind and the finishers had not been able to remove them. The article warned against having any loose insulation of this kind in areas where it could become windborne on the day that flatwork is being done.

I was recently involved with a floor where this had occurred and where the beads were effectively removed by dissolving them with a solvent. Then, to fill the voids, a squeegee application of a 100-percent-solids epoxy coating can be applied, followed by a roller application of the same material. This method should solve the problem if the slab is a floor. On exterior slabs in colder climates an epoxy should not be applied but a breathing type of coating or topping should be used instead.

Rained-on slab may not require repair

Q. *We finished an air-entrained concrete ramp at a truck loading dock and left it uncovered late one evening. It rained shortly afterward and surface mortar was washed off. We didn't try to refinish the surface then. What can be done to repair it now?*

A. Repair may not be necessary. Because you didn't attempt to refinish the surface, which would have worked water back into it, the only damage is loss of surface mortar. The slab now has an unplanned exposed aggregate surface. Because it's air-entrained and sloped to drain well, the slab will probably resist freezing and thawing cycles. Repair would ordinarily involve light grinding and scoring to improve traction, but the grainy texture you now have should be adequate for traction. Unless appearance is a problem, which seems unlikely for a loading dock, you may have produced exactly what the owner needs.

Crack repair

Temporary repair with epoxy

Q. *I am a structural engineer and am investigating a problem with a parking garage below grade. This garage is part of a larger building and projects beyond it. It has a two-way flat slab at grade like a bridge. It is about 6 years old. There are cracks over the large rebars, some of which are #10 bars. The drawings show only 1/2 inch of cover over the steel, and some rusting is visible. There is cracking in the slab, mostly over the columns. The owner is not prepared to do major structural repairs immediately. What I would like to know is whether we could inject the cracks with an epoxy and get 4 or 5 more years out of the garage before undertaking major structural work? We would monitor the performance in the meantime.*

A. The fact that you plan to monitor the performance makes the plan sound safe. That makes the answer boil down to one of economics. An epoxy properly applied should remedy the existing problem though it would not prevent the continuing deterioration associated with too little cover and too much deicer salt probably already in the slab.

Should driveway cracks be repaired?

Q. *My company does concrete sidewalk and driveway work. My crews put in joints that are deep enough and close enough together to control cracking, but occasionally we get a random crack. The owner always wants me to fix the crack. Is there any way to glue a cracked driveway back together so the repair isn't noticeable?*

A. You can blend epoxies to closely match the color of in-place concrete, then inject the material into the crack. But it's usually not worth the effort and expense to do this for a driveway. More often, a standard gray epoxy is used, cleanup isn't too good, and the repair is more obvious than the original crack.

For a slab on grade, a random crack isn't likely to affect slab performance any more than a planned joint would. The crack just looks bad. If you've placed the slab on a well-compacted base and used good quality concrete, it should still perform well. Sealing the crack will keep out water that might soften the subgrade, but the sealing material will also make the crack more noticeable.

Choosing a crack repair method

Q. *We've recently undertaken a few crack repair jobs and have become acquainted with a variety of methods, including epoxy injection, routing out and resealing, and using overlays. Now we're thinking of going more into the business but would like to be sure we understand the best approaches for various situations. Looking over some of the literature, we find the number of approaches bewildering. Has anyone done a systematic study of the many methods available and published a comparison?*

A. In the May-June 1984 issue of the *ACI Journal,* beginning on page 211, there is a 20-page report by ACI Committee 224 called "Causes, Evaluation, and Repair of

Cracks in Concrete Structures" that covers the subject well. The same report can be found in the *ACI Manual of Concrete Practice.* A condensed version of about 5 pages, which doesn't include the information on causes of cracking or evaluation of their condition, was published in *Concrete Construction* in January 1985, page 37, under the title "How to Fix Cracks." These articles give a pretty good overview of the subject.

Sealing cracks in post-tensioned deck

Q. *Recently we completed construction of a hotel parking garage. The lobby level of the garage has a 4-inch topping on a 10-inch post-tensioned slab with post-tensioned beams spaced at 27 feet. The structural slab was not waterproofed before putting on the topping. Because of heavy traffic the topping slab is now cracking and allowing water to pass through and become trapped above the structural slab. Since there is no other escape, this water is being forced through small shrinkage cracks in the structural slab, as evidenced by buildup of calcium hydroxide on the bottom of the slab. Is there some way to inject a waterproofing material into these cracks to prevent water penetration?*

A. We have no definite solution but offer the following comment. If there aren't many cracks in the post-tensioned slab and you have good reason to believe that there won't be many more, you might successfully seal them from the bottom by injecting with epoxy resin, as you suggest. You might wish to talk about this with a company with long experience in the field.

There are also some polyurethane foam grouts that can be injected into moisture-containing cracks to seal them. These are described in the article "An Effective Repair for Leaking Waterstops" published in our June 1984 issue, page 594.

If you have many cracks in the post-tensioned slab or have reason to believe that cracks will continue to develop in it, you may want to consider some means of sealing the top surface of the topping to keep water from penetrating it and ultimately the post-tensioned slab as well. There are some elastomeric coatings or membranes that resist wear from traffic and have been used on parking deck slabs. They're applied in thickness of about 60 mils.

If you apply such a coating or membrane over the top it may take awhile before the existing water gets out of the slab. For that reason you may want to devise a way of checking periodically to see that the water in the slab is not corroding the post-tensioning cables.

Two other articles that contain information on sealing parking deck slabs are: "Parking Deck Deterioration—the Reasons Why," in our September 1976 issue, page 444, and a short item "High Molecular Weight Methacrylate Used as a Crack Welder," in our October 1984 issue, page 885. The first discusses several approaches to waterproofing a parking deck. The second discusses the squeegee application of liquid methacrylate to the top surface of a bridge deck to seal cracks.

Repairing faulted joints

Q. *What is the best way to repair joints in pavement in which the slab on one side of the joint may be from $^1/_8$ to $^3/_8$ inch lower than the adjacent slab?*

A. First the condition of the concrete and subbase in the vicinity of the joint must be determined. The slab must be stabilized as needed in this area by pressure grouting and installing edge drains. Spalls must be repaired and joints resawed and sealed.

The two slabs can then be brought to the same level by a diamond grinding machine that cuts the surface of the high slab down to the level of the low slab. The American Concrete Pavement Association says that a fault $^1/_3$ inch high may be tapered to zero in a distance of 3 feet from the joint, while a $^1/_2$-inch fault may require a distance of 5 feet. Grinding is done with a multiple-blade saw that produces skid-resistant corduroy grooves in the longitudinal direction of the pavement. The work can begin at either the high or low end of the portion to be cut. Residue from the grinding operation should be removed so that it does not flow into gutters and drains or become airborne through the action of traffic or wind. Four-lane pavements may require grinding of only the slow-moving lanes, since faulting may be negligible in the lanes that have carried little truck traffic, though some feather-edge grinding may be needed where the two kinds of lanes abut.

Bulging block wall below grade

Q. *One wall of our block basement is bulging inward and it has cracked horizontally for the length of the wall. We don't have any accurate measurements, but we think it has moved inward at the center about an inch during the past week. The wall is about 25 feet long and we have been having repeated trouble with leakage on that side of the basement, though a waterproofing company had some time ago introduced a drainage system on the inside at the level of the footings. This keeps getting clogged up and the company keeps coming back to clear it out. What can we do about a permanent remedy?*

A. You should perhaps have a structural engineer look at the wall and evaluate the situation. The wall very likely needs to be reinforced. A good method is described in our February 1983 issue, page 141. After the wall has been reinforced, you should perhaps add some fill outside the wall to drain surface water away from the house at grade level. This should reduce the load on your subgrade drainage so that it doesn't continually plug up.

Remove and replace or top with unbonded overlay?

Q. *Our firm was hired to repair leaks in the concrete roof of a college building in North Carolina. The roof doesn't have many cracks, but most of the joints are leaking where PVC (polyvinyl chloride) waterstops placed in the joints have separated from the concrete. We're seating the waterstops in epoxy, rebuilding the joint faces, placing foam backup rods in the joints and then sealing them with an elastomeric sealant. Because the roof deck is regularly used by pedestrians, we were also asked to restore the deteriorated concrete surface. With two scabblers we spent a week removing the top 1/4 inch of concrete, but the deck still produced a hollow sound when we pulled a chain across it afterwards. We think it would take too long to remove the entire 4-inch surface slab with a scabbler or a planer. With a jackhammer, we're afraid we'd damage the waterproofing membrane between the bottom structural slab and the top surface slab. Are there any other ways of removing concrete that would be practical?*

Instead of removing the old concrete we're also considering placing an unbonded concrete overlay reinforced with polypropylene fibers. We can increase the elevation 1 to 2 inches without problems, but can an unbonded overlay be placed this thin? Should we construct joints in the overlay closer together than the 24-foot joint spacing of the existing slab, so that flexural cracks don't occur? What materials can we use to break bond?

A. Scarifiers or water blasting equipment are other tools that might be used to remove the surface of a concrete slab to a specified depth. Manufacturers of such equipment are listed under the headings, "Scarifying Machines" and "Scarifying Tools," in the December Buyers' Guide of *Concrete Construction.*

According to ACI 302.1R-80, *Guide for Concrete Floor and Slab Construction,* "[an unbonded] topping must be a minimum of 2 1/2 inches thick, but thicker toppings are recommended." Thus you ought to devise some way of accommodating a topping 2 1/2 inches thick. Plastic sheet, felt paper, a layer of sand or a chemical bond breaker can be used to prevent bond to the base slab. Manufacturers of chemical bond breakers are listed in the December Buyers' Guide of *Concrete Construction* under "Bond Breakers." To control cracking in slabs, American Concrete Institute Committee 302 has recently agreed to adopt a new rule of thumb in place of fixed joint spacings. This rule is that the joint spacing should be twice the distance in feet of the thickness in inches. Thus a 2 1/2-inch topping would require joints every 5 feet in both directions. Since 5-foot spacings wouldn't fit well into the previous 24-foot spacing pattern, you would probably want to reduce the spacings to 4 feet.

Repair of crack in topping over precast planks

Q. *I have a 24- by 26-foot slab 3 inches thick cast over precast concrete planks on top of 8-foot walls. It has been in for 3 years. It has a full-length crack over the center seam of the planks. How can I fix it?*

A. Such a crack is caused by drying shrinkage, thermal contraction, differential movement or a combination of these. Probably the best way to repair it would be to create a joint by making two parallel saw cuts the length of the slab, one on each side of the crack, and then breaking out the concrete between them to a depth of about 1/2 inch. The depth should be about half the distance between the two saw cuts. The joint should then be sealed with an elastomeric joint sealant such as polyurethane. The polyurethane should be prevented from bonding to the bottom of the joint by putting in a strip of paper, polyethylene or roofing felt so that the joint sealant can stretch freely.

Patching and Topping

Is epoxy coating needed for rebar in latex modified concrete?

Q. *The engineer for our parking garage repair job required us to paint an epoxy coating on some of the exposed bars before placing a latex modified concrete overlay. Is this really necessary?*

A. The suppliers of latex modifiers assure us that a properly installed overlay of latex modified concrete effectively protects against corrosion and that the epoxy is not necessary. They seem to have enough field experience to support their claims. However, designers of parking structures, particularly in municipal work where maintenance budgets are limited, frequently take a "belt and suspenders" approach. At times they also specify corrosion-inhibiting admixtures in the concrete surrounding epoxy coated bars.

Remember, for an effective repair job the overlay must be well bonded to the old concrete. This means that its surface must be absolutely clean and free of foreign matter, including particles of epoxy. Carefully clean up any spills or dripping of the epoxy from the rebars onto the old concrete before placing the overlay.

Pier repair

Q. *We have a project consisting of repair of a pier on the Atlantic Ocean in a subtropical climate. The job consists of repairing the original concrete piles that have been damaged not only by the effect of the seawater but by mechanical abrasion due to ships. What procedure should be used?*

A. The piles can be repaired by removing loose and unsound concrete and then casting new concrete around them within special jackets.

One kind of jacket is a nylon fabric sleeve that surrounds the pile and is closed with a vertical zipper.

Another kind of jacket is made of a fiberglass-reinforced plastic sheet that can be bent into a cylindrical shape and held in place by an interlocking closure that is part of the sheet. Special jackets are also available with rectangular cross-sections and several kinds of accessories. Concrete or mortar is pumped in through a valve at the bottom.

Another solution is to repair by shotcreting. After removing unsound concrete, steel mesh or anchors should be attached to the piles. Shotcreting would have to be done at low tide. The shotcrete layer should be thick enough to provide 2 inches of cover over the steel.

Repairing piles in seawater

Q. *The piles of an ocean pier have deteriorated to the point where they have to be repaired soon, and some of the reinforcing steel will have to be replaced. The thickness of the repair will vary from about 1 1/2 to 3 inches and we would like a recommendation of a good method of forming and concreting.*

A. Why not use a fabric pile jacket? After all the chipping has been completed and the necessary new steel has been anchored in position, the fabric jacket is hung around it. This is a sleeve tailor-made for the pile and it is provided with a zipper closure. It hangs from a ring in a hem at the top of the sleeve. When the zipper has been closed the bottom is bonded to the pile and the top ring is pulled upward with turnbuckles until it is tight. The repair is made by injecting mortar through two flexible hoses that reach to the bottom of the sleeve on opposite sides of the pile. These hoses are withdrawn as the mortar fills the sleeve.

More on repairing piles in seawater

Two readers of the Problem Clinic item "Repairing Piles in Seawater," October 1981, page 849, have called our attention to methods of repair other than the method we described. The two other methods are described below.

1. The first method was successfully used about five years ago on more than 1500 LaGuardia Airport runway piles in Riker's Island Channel. After the deteriorated or damaged pile surfaces have been properly prepared, a crew approaches each pile in a raft or boat at low tide. The raft or boat contains a small mixer and supplies of a two-component epoxy and filler designed for this particular purpose. After a batch is mixed a workman wearing rubber gloves picks up baseball-size globs of the mixture and kneads them onto the piling by hand. It is said that a pile length of about 6 feet can be coated to a thickness of $^3/_{16}$ inch in about seven minutes after the crew has become experienced. The epoxy to be used for such a surface should be chosen only on the basis of tests simulating the service conditions to be expected.

2. The second method makes use of a fiber jacket or fiberglass-reinforced plastic jacket around the pile which serves as a form face during placement of concrete. If a fiberglass-reinforced plastic jacket is used it is left in place to provide strong, rigid protection against erosion, wave action, freezing and thawing, marine-bore attack and chemical attack and to resist impact damage. This kind of jacket also provides a straight, uniform, attractive appearance. Fiberglass-reinforced jackets have been in use for several years and are reported to have gained widespread acceptance by state departments of transportation, port authorities, the United States Navy and the Corps of Engineers. Repairs with them have been accomplished to the mud line in depths up to 60 feet, withstanding pump pressure up to 8600 psf. The success of the fiberglass-reinforced jackets is reported to have been such that the system is now being specified to protect piling in new installations.

Repair of face of spandrel beam

Q. *A spandrel beam on the front face of a two-year-old shopping center is not plumb. The beam is 4 to 5 feet wide and 50 feet long. Some time ago it was given a patch that varied in thickness from 1 inch at the top to nothing at the bottom. The patch includes some metal lath. Water has made its way into the patch and caused corrosion and much of the patch has sheared off.*

We must repair the beam again. There are no shotcrete operators in our area. Should we use a latex bonding agent? Should we featheredge? Should we apply a curing compound? The repaired beam will finally be painted.

A. We suggest that you prepare the surface well by acid etching by recommended procedures including thorough rinsing. Then scrub in a 1:1 cement-sand grout made to a creamy consistency. Immediately, before the grout can dry, apply a portland cement plaster mix containing a latex. There are proprietary mixes of this kind on the market known as finishing mixes. This mix should trowel readily and doesn't necessarily need a curing compound, though a curing compound should help. At some appropriate point in this work you should investigate all the joints above the spandrel beam. One or more must be leaking and any leaking joints should be resealed. The joint above the beam should preferably be flashed using a flashing with an unobtrusive drip ledge.

Repairing effloresced pier caps

Q. *We have a project to replace an old concrete deck. The existing pier caps are heavily covered with efflorescence. Your "Guide to Repair of Concrete," March 1977 issue, gives good descriptions (pages 124 and 131) of how to remove efflorescence. But you do not mention how to test to determine whether or not the concrete is still in good condition.*

A. The application of muriatic acid to remove efflorescence from a bridge pier cap is not likely to cause significant loss in structural strength unless there has been some unusual condition or excessive application of acid. Under ordinary conditions the acid washing should affect only about the top 1/8 inch of surface. If there are cracks in the pier cap the acid could penetrate more deeply and attack reinforcing steel, but even so the amount of acid ordinarily used would not be likely to cause significant damage provided the cracks were then repaired to prevent subsequent intrusion of water.

If you have reason to want to check the strength and integrity of the pier cap you could perhaps get the best results by the "Standard Test Method for Pulse Velocity Through Concrete," ASTM C 597. The test should be performed by someone who is thoroughly familiar with its use and interpretation. It will give information on the quality and uniformity of the concrete, which can then be compared with that of nearby concrete of the same mix design and age which is known to be good. The test will also detect internal cracks or other discontinuities.

The relative strength can alternatively be checked by "Tentative Test Method for Rebound Number of Hardened Concrete," ASTM C 805, or "Tentative Test Method for Penetration Resistance of Hardened Concrete," ASTM C 803. As in the pulse velocity method, these results also would have to be compared with those for nearby concrete.

A destructive method of test is to take drilled cores from areas suspected of being weak or damaged. This is done according to "Standard Method of Obtaining and Testing Drilled Cores and Sawed Beams of Concrete," ASTM C 42. The cores can be tested directly for compressive strength and examined visually for cracks or other obvious troubles, and, if desired, can be examined petrographically to check for any other kind of deterioration.

Rebars need care during patching

Q. *In patching work on bridge decks what should be done about rebars that are cut, bent, damaged by the work in progress, or corroded?*

A. First of all, work around rebars should be done with care to avoid unnecessary damage from hammers, chisels or other tools. If any damage is done to a rebar, or if it has been corroded enough to reduce the cross-sectional area, a length of new rebar should be inserted parallel to it. Any exposed rebars that are higher than the level of surrounding bars should be bent back down so that they will receive an adequate amount of concrete cover when patching is done.

Patching board-finished concrete

Q. *Our company placed a poured concrete wall with a board finish. There are a few honeycombed areas that have to be patched. How can we make the patches as unnoticeable as possible?*

A. It's nearly impossible to make a patch that will blend perfectly with the rest of the concrete. If the honeycombed area isn't too large, it will probably be less noticeable left as is. A poorly matched patch may draw more attention than the original defect (see photos next page).

If the owner insists on patching the honeycomb, first work on matching colors of the repair material and existing concrete. Patching mixes for shallow surface voids are usually mortars. As a starting point in designing the mortar, use the cement-sand ratio from the actual project design mix. For example, if the design mix calls for 564 pounds of cement and 1,190 pounds of sand per cubic yard, the cement proportion is 32% of the dry ingredient weight.

A batch made with 5 pounds of dry ingredients is about the right size to permit placing before the mix stiffens too much. For a batch of this size containing 32% cement you'll need 1.6 pounds of cement and 3.4 pounds of sand.

Starting with a known amount of water in a container, add water to the mix until it reaches a stiff ball consistency. Sometimes it helps to add an acrylic or latex bonding agent. Use about 10 to 20 milliliters per 5-pound batch.

After mixing in the bonding agent, use the trowel-slump test to measure batch wetness. Pick up mortar on the back side of the trowel and tap the trowel on the side of the mixing bowl twice. Then turn the trowel so the blade is vertical. If the mortar *slips* off it is too wet and the batch should be discarded. If the mortar *falls* off it is too dry. Add water as needed to get the right consistency.

Compute the final amount of water used by subtracting the amount left from the amount you started with. Record all batch weights so you can duplicate the batch.

To test the color match, try placing some of the mortar in a part of the structure that won't be seen. Let it cure for several days by first spraying the patch with

Reasonably good patching of board-marked concrete. The board edge lines have been carefully restored for continuity through the repaired area. If the color match had been better, this repaired area would go almost undetected.

PHOTOS BY: © Richard W. Steiger

An aesthetic failure in patching, although probably serviceable and durable. Apparently no thought was given to either color or texture, leaving an unsightly blemish that detracts from the architectural finish.

water, then covering it with polyethylene held by duct tape. Color of the patch should be slightly on the dark side since it will bleach out with age.

Don't repair outdoor plastic shrinkage cracks

Q. *We recently installed a concrete walk that was damp cured. We have noticed thin but deep cracks in a few spots. The contractor called them "heat cracks." Though they are not wider than one's fingernail I am concerned that they will worsen as the freeze-thaw season approaches. Is there available an epoxy or something that I can use to seal these cracks? There is no cure-and-seal membrane to interfere with anything we apply.*

A. We gather that these are short cracks that begin and end somewhere in the body of the concrete—not cracks that begin at the edge and run for some distance or all the way across. If so, they are perhaps what are usually known as plastic shrinkage cracks.

Plastic shrinkage cracks occur when the rate of evaporation is high during construction. Conditions that contribute to rapid evaporation are high temperature, low humidity, wind, a slow rate of bleeding or some combination of these. On hot, dry windy days, or at other times when plastic cracking might be threatening, a variety of measures can be taken to minimize rapid evaporation. These include using sunshades and windbreaks, dampening the subgrade and formwork, and initiating the curing process promptly.

Your problem, however, is what to do about the cracks. We have recently looked at some plastic shrinkage cracks in a street in Wauwatosa, Wisconsin (a Milwaukee suburb) that developed when the street was built about 36 years ago. These cracks were considerably wider than yours and there was public concern at the time over what was perceived as a deficiency in the new concrete. However, the city was persuaded that the cracks would do no harm as far as service was concerned, so they made no repairs at the time and the pavement is still in otherwise good condition even though the street has been routinely salted in winter. The main difference is that the cracks have become filled with dirt and so they're harder to find. The appearance has also changed slightly because of some surface wear. This street has been discussed in the article "Common Field Problems," October 1985, page 855.

It is possible to seal the cracks by some such means as epoxy injection. But the usual result is to accentuate their presence. If the cracks are in fact plastic shrinkage cracks, the chances are that they won't get any worse and that the best thing you can do for their appearance is to leave them as they are. If the concrete is properly air entrained it should survive cyclic freezing and thawing.

Slab over a damaged slab

Q. *We have been hired to place a 4-inch-thick concrete slab over an existing one. We plan to break up the old slab with an iron ball. However, one area of the old slab has heaved during the winter, probably because water has gotten underneath the slab and frozen there. Do you think we should break up the old slab as planned, and if so what precautions should we take in the area that is heaving?*

A. If built on a very rough base of broken concrete, the new concrete slab would undoubtedly crack. We suggest you don't break up the old concrete unless you plan to remove it. You need to do something to prevent the cracks in the base slab from being transmitted into the new slab. You could put down a thin layer of sand, 1/4 to 1/2 inch thick, or you could apply a layer of polyethylene sheeting to break the bond between the new and the old concrete. In either case, you should take care while placing the new slab to avoid tearing the polyethylene sheeting or disturbing the sand layer so that you don't allow fresh concrete to bond to the old slab. You should remember that concrete placed on top of polyethylene sheeting will usually produce more bleed water and this should be allowed to evaporate before doing any finishing or troweling.

The area that has heaved should be broken out, and the earth removed and replaced with crushed stone, gravel, or some other material that will not undergo expansion during freezing weather. This subbase should then be properly compacted before concreting.

Unbonded topping

Q. *There is a floor in an old building which we intend to convert to a warehouse. There are some differences in floor surface elevation due to random cracking and settlement. We want to put in a new floor. Since increase in elevation would be no problem can we simply cast the new floor over the top of the old one? Would this be a place where we could use an unbonded topping? If so, would a layer of sand be suitable to prevent bonding?*

A. Yes, you could cast an unbonded topping of 2 1/2 inches or greater thickness over the old floor. A sand layer would be a suitable bond breaker provided it were of sufficient thickness to cover the highest areas of the old slab and were uniformly compacted. It would be wise to study the recommendations of "Guide for Concrete Floor and Slab Construction," ACI 302.1R with respect to unbonded toppings.

Heaving floor in basement

Q. *The concrete floor in my brother's basement has cracks along both sides and both ends, about 3 feet out from the walls. There are also some diagonal cracks running from these cracks to the corners. There aren't any other cracks in the floor. The floor also seems to bulge toward the center. There would be room to put a topping over the whole surface to level it up and cover the cracks. Is this a good solution, and if so what would be the minimum thickness needed at the highest point?*

A. What you describe sounds like a floor being heaved by an expansive soil. To confirm this you might look for other signs or clues: outside ground sloping toward basement walls, hardly any cracks in the foundation walls, opening in the living area out of level, floor in the living area out of level, doors binding and hard to open or close, and cracks developing around door and window areas in the living area. If you brother didn't notice any trouble at all for the first few years, if he has the troubles just enumerated and if these troubles are getting worse, his main problem is probably an expansive soil.

If so, you should consult an expert of repair of concrete on expansive soils. Heaving of the soil is caused by buildup of water around the base of the foundation, so the first thing to do is to stop this process. The grade of the yard should be corrected to slope at least $^1/_2$ inch per foot away from the foundation to promote drainage of rainwater. It will also be necessary to prevent formation of channels through the soil to the foundation. These form when the soil dries and cracks. Hence there is a paradox: to prevent water from reaching the foundation, keep the soil from cracking by watering it consistently during the dry periods of the year. (In a soil that is already cracked, water a little at a time until the cracks are finally closed.) These measures will allow abatement of heaving and prevent further damage to the foundation walls and the house framing.

The next step is to cut the floor free of any attachment to foundation walls. This job must be done by someone with experience with such floors because there is a possible danger from a sudden release of potential energy. The floor should also be freed from attachment to plumbing stacks and—though you didn't mention them—any columns or fireplace masonry.

After the cause of heaving has been removed, a good way to repair the floor is to cut it along lines about 12 inches on each side of the cracks and remove the resulting 24-inch wide strips of concrete. Then fresh concrete can be placed in the same area and sloped from the high side to the low side, thus getting rid of the abrupt change in level. If this kind of repair is not to your liking, then a completely level surface can be obtained by placing a concrete topping, at least 2 inches thick at the highest point, over the whole floor.

Before undertaking any repair of the basement floor it is a good idea to trowel a thin layer of plaster of paris over a number of short lengths of the crack to see if the floor is moving. If the plaster of paris finally cracks, add another series of patches. When a set of patches do not crack during a period of 3 or 4 months, or just barely crack, it can be assumed that heaving has been arrested and it is safe to repair the floor.

Don't bushhammer gunite substrate

Q. *We're required to remove, clean up, and replace gunite that we recently applied to cover a concrete wall surface that was spalling. We had faithfully followed specifications and took special care with the bushhammering of the old surface to be sure that it was all thoroughly roughened. Much of the gunite came off within a month after it was applied. We have not had this trouble before and we wonder if you have any idea of what might be wrong.*

A. Bushhammering should not be done to prepare a surface for guniting. Although it does roughen the base, or substrate, bushhammering inhibits bond because it powders both the cement paste and rock and drives the fine particles into the surface pores. These particles prevent the gunite itself from entering the pores and bonding thoroughly to the old concrete, or substrate. This leaves a weakened plane between the new material and the old and encourages irregularly shaped slabs of the new material to come off later.

If bushhammering was required by the specifications, you should get the specifications changed before applying new gunite. The surface should be prepared by chipping off the gunite you applied and then sandblasting to remove the powder produced by bushhammering. The substrate should then be thoroughly moistened to prevent it from soaking up water from your new shotcrete.

Early patching of tie holes

Q. *The owner's representative for a large job insists that the contractor patch the tie holes as soon as he strips the forms. (He has been keeping the forms heated and then stripping them after 48 to 72 hours.) As consultants, we prefer to wait until warm weather to patch the holes, then apply a grout cleandown finish. Is it necessary to patch the holes immediately?*

A. It shouldn't be necessary. If it is intended, however, that the pattern of patches should not be visible it will be necessary to take whatever measures are required to see that the texture and porosity of the tie hole patches match the rest of the wall. This will prevent them from showing through the grout. If, when it is time to apply the grout finish, the tie hole patches are not at about the same moisture content as the rest of the wall, it would be well to moisten the wall uniformly just before applying the grout.

Prepare well for patching

Q. *Experience shows that repairs have been made at disturbingly frequent intervals on a bridge deck that bears heavy traffic. Many of these repairs are replacements of patches that have been put in more than once before. The patches have been made of good quality concrete containing a latex and carefully featheredged. Old concrete has always been removed right down to the rebar. Can you suggest a better patching method?*

A. Your trouble may be in getting the right configuration for the patch. Sufficient deteriorated material, whether old patching mix or original concrete, must be removed to give at least 3/4 to 1 inch clearance around all exposed rebars, and the bars should be thoroughly cleaned by sandblasting to remove thick, scaly rust, though not thin, tightly bonded rust. If the bars are deteriorated, a structural engineer should be consulted about how much lap to provide on the new steel that you attach to the existing bars.

In cutting out the concrete, the edges should be made sharp, not featheredged, though corners should be rounded. It is also a good idea to undercut the edges so the patch will be wider at the bottom than at the top, providing keying action. Before installing the new patch, all the surfaces should be thoroughly cleaned to remove loose particles and dust. This can be done by high pressure water jetting, sandblasting, blowing with compressed air, or vacuuming. If the exposed surfaces contain oil, asphalt, or other contaminants, those also should be removed.

Temperature extremes

Floor to resist molten metal

Q. *We have been engaged to replace a floor in a brass foundry which has disintegrated from spillage of hot molten brass from ladles that move across the floor between the melting furnaces and the casting area. What is the best way to install a floor that will provide maximum resistance to continuing spillage of this kind?*

A. A refractory concrete might be able to survive such high-temperature abuse. We suggest investigating the use of a calcium aluminate cement. The old floor should be taken out completely and a full-depth floor built with concrete made of the special cement and a refractory aggregate.

Heat-resistant heavy-duty topping

As an elaboration on the Problem Clinic item, "Concrete to Withstand Hot Metal Spillage," (January 1981, page 122), Guy H. Wilson, Special Products, Tews Lime and Cement Company, Milwaukee, has sent the following procedure for applying a heat-resistant heavy-duty topping for concrete floors.

Start with a clean subbase from which all laitance has been removed. Prepare a bonding grout composed of one part aluminous cement, one part fine emery and enough water to produce the consistency of heavy cream. Predampen the base and apply the bonding grout.

The topping should be placed immediately, before the bonding grout has had a chance to dry. This topping should be proportioned 94 pounds of aluminous cement to 300 pounds of emery graded from 3/8 inch to fines. Potable water should be used up to a maximum of 3 1/2 gallons per 94-pound bag of aluminous cement (w/c = 0.31).

This should produce an extremely dry topping that will require placement with a tamper or roller followed by screeding to final level. Float with a compactor-type power float and trowel to the desired finish. If there are joints in the subbase, joints should be installed in the topping directly above those in the subbase. Curing is extremely important. Use only water, either by spraying at frequent intervals or ponding for a minimum of 24 hours. Do not cure with membrane curing compounds.

A rule of thumb is to provide 1 1/2 inch of thickness per 1000° F plus 1/2 inch for each additional 500° F. As the melting point of aluminum is 1220° F, a 2-inch-thick topping would be adequate for the floor described in the previous Problem Clinic item.

Guaranteeing durability

Q. *Can concrete be guaranteed for outdoor service in a cold climate where freezing and thawing occur, and if so for how long?*

A. If you are thinking about concrete in general, the answer is No. Some concrete will last for years and years, if not centuries, and other concrete will be disrupted fairly soon. If you are thinking of air-entrained concrete, the answer is Maybe. It depends on whether the concrete was properly proportioned, batched, mixed, placed, finished, protected from the weather if necessary during curing, and properly cured. If you are thinking of air-entrained concrete placed by an experienced contractor who takes all the necessary precautions to use concrete adequate for the purpose and to perform all operations properly, the answer is Yes.

In general, the main consideration in producing concrete that will withstand numerous cycles of freezing and thawing is to use a mix with an adequate air content. ACI 301 "Specifications for Structural Concrete for Buildings," makes recommendations for the amount of entrained air depending on the maximum size of coarse aggregate. But it is not enough to rely entirely on air content if the water-cement ratio is high, if low-grade aggregates are used, or if good job practices are not followed.

As for how long concrete that is properly designed and built can be expected to last in a freezing and thawing environment, it is hard to say. The first concrete street built in this country has survived the freeze-thaw environment of Bellefontaine, Ohio since 1891. This street was not made of air-entrained concrete, but it had such a low water-cement ratio that it has survived anyway.

It was in full use until 1960 when obsolescence and economics dictated covering parts of it with a topping.

Freeze-thaw durability of shotcrete

Q. *We have been building concrete arch culverts with shotcrete. They're generally protected by earth cover, but we're wondering about durability at the water line. How well will they resist freezing and thawing damage? Should we use air-entrained shotcrete?*

A. The American Concrete Institute's "Recommended Practice for Shotcreting" (ACI 506-66: revised 1983) cautions that it is not feasible to entrain air in dry-mix shotcrete. However, air-entraining admixtures may be used in many wet-mix applications. To entrain a given percent of air in the shotcrete as placed, ACI 506 recommends about twice that quantity in the original mix to allow for loss of air in gunning.

According to ACI 506, data on the effect of entrained air in shotcrete are limited, but air entrainment would be expected to increase resistance to freezing and thawing damage and to make some mixes more workable.

We received the results of some high-strength shotcrete tests from Rusty Morgan of Hardy BBT Limited, Burnaby, British Columbia. One objective of the tests was to study freeze-thaw durability of both wet and dry mixes, and the results are interesting although the number of specimens was limited. Morgan tested three samples from each of six mixes under study:

WET MIXES

- Plain control
- With 13.1% silica fume (microsilica)
- With silica fume and steel fibers

DRY MIXES

- Plain control
- With 12.6% silica fume
- With silica fume, steel fibers, and accelerator

All the wet mixes also contained an air-entraining agent. The dry mix was designed with a water-cement ratio of 0.29 and the wet mix, 0.35. Their 28-day compressive strengths were in the 9000-12,000-psi range.

After 300 freeze-thaw cycles, the study reports that excellent durability factors above 96% were achieved with all of the mixes tested, whether containing an air-entraining agent or not. Spacing factors for the air voids in all the mixes were significantly higher than the 0.008 inch (0.2 mm) generally recommended for air-entrained concrete. Tests were conducted in accordance with ASTM C 666, "Test for Resistance of Concrete to Rapid Freezing and Thawing," Procedure A, with one exception. Longer periods of moist curing of the specimens were used to permit most of the pozzolanic reaction with the silica fume to take place before the freeze-thaw cycling. Measurements of the air void system were determined using ASTM C 457-82 "Microscopical Determination of Air-Void Content and Parameters of the Air-Void System in Hardened Concrete," modified point count method. Compressive strengths of all the shotcrete mixes studied were higher after undergoing the freeze-thaw tests; the increase ranged from 10% to 65%. Overall, these results suggest high durability of high-strength shotcretes, regardless of air-entrainment. A complete report of these tests has been submitted to a technical society for publication, and *Concrete Construction* will advise readers when it is available.

It would still seem prudent to use an air-entraining agent for lower strength shotcretes where exposure to freezing and thawing is expected. According to Schrader and Kaden,* "The high-quality paste and mortar achievable with shotcrete having low water-cement ratios, low porosity, and relatively high-strength results in a higher initial resistance to damage, but it will still deteriorate progressively and extensively when saturated and frozen if it has strengths less than about 10,000 psi, and if it does not contain proper entrained air bubbles." However, they mention the possible effectiveness of latex modifiers in shotcrete that is not air entrained. "Because properly formulated and applied latex modified shotcrete has extremely low porosity and permeability, and because of the apparent reduction in interconnecting capillaries, it is virtually impossible to saturate it sufficiently to cause freeze-thaw damage. It is virtually watertight and may become vaporproof."

---

*Schrader, E., and Kaden, R., "Durability of Shotcrete," pages 1071-1101, *Concrete Durability,* ACI Special Publication SP-100, American Concrete Institute, Box 19150, Detroit, Michigan 48219, 1987.

Too much heat

Q. *I have been asked to repair an area of about 10 feet by 10 feet in a 6-inch-thick floor on grade. This floor, in a building used by an oven manufacturer, had performed well for 15 years. Recently the company tested out a new experimental oven. I believe the oven did not have a bottom, but they set it on some insulation and heated the oven to 1100° F. After some period of time the whole 10x10-foot area exploded. Concrete came off for depths* *ranging from $2^1/_2$ to $3^1/_2$ inches. None of the rubble from the explosion was larger than $^3/_4$ inch diameter. I expect to patch this by sawing out the remainder and replacing the area to the full depth, but I am curious about what caused the concrete to blow apart.*

A. The behavior of the concrete was similar to the way concrete performs in a standard fire test. The insulation, incidentally, does not prevent heat from going into the concrete; it just reduces by a small amount the maximum temperature in the concrete.

If the room temperature is 70° F and we overlook the effect of insulation, the temperature rise is 1100 minus 70 or 1030° F. The concrete in the heated area tries to expand greatly, because of the higher temperature. This potential expansion is almost completely restrained, however, by the large surrounding area of unheated concrete. Multiplying the temperature change by the coefficient of thermal expansion, 0.0000065, gives 0.0067 inch per inch potential expansion.

Assume this potential expansion were completely restrained by the very large area of surrounding cool concrete, and that the concrete had a modulus of elasticity of 4,000,000 psi. (In a 15-year-old slab on grade, which would probably be fairly moist, the value could be even higher.) The maximum compressive stress would be 0.0067 x 4,000,000 = 27,000 psi at the top surface. This stress diminishes rapidly from the hot surface to the cooler depths of the slab, but the gradient in compressive stress causes great shear stresses. Together the compressive and shear stresses destroy the affected layer of the slab.

There are a considerable number of other factors that are thought to contribute to what happened, but this is perhaps the main cause.

Vacuum dewatered vertical surface resists heat

Q. *Can a concrete wall be made more resistant to intense heat by vacumm dewatering? If so, could ordinary concrete be used to build the walls and floor of a soaking pit for steel ingots? Has anyone worked out the methods?*

A. Suitable techniques are available but you will have to be the judge of whether the concrete would have suitable heat resistance for your purpose. An example of the use of vacuum dewatered concrete for heat resistance is the 8-inch-thick faces of the ducts under the space shuttle launch pad at Vandenberg Air Force Base in California. The walls and floor of these ducts are exposed directly or indirectly to the blast of the rocket flames when the shuttle lifts off.

The method used was to build the flame ducts—or tunnels—of reinforced concrete 6 to 9 feet thick. The interiors of the ducts are 50 feet high, 70 feet wide, and 330 feet long. The walls and floors of the first 100 feet of each duct were covered with a layer of vacuum dewatered concrete made with Type II cement—8-inches thick on the floor and 6 inches on the walls.

The floors were dewatered by the standard techniques. To dewater the walls, 10-foot-high forms were first built for use in placing the 8-inch-thick facing. A hole was drilled in the center of each form to connect to the vacuum manifold. A self-adhering rubber gasket was placed around the perimeter of the form face to seal it against the concrete, and vacuum filter pads were stapled to the form within the gasketed area. A nylon mesh was then stapled over the filter pads, and a silk cloth, soaking wet, then stapled over the nylon mesh. The assembly was set aside to dry 24 hours, leaving a taut silk surface to contact the concrete.

Forms were tightly butted, 6 inches out from the existing wall, and concrete was placed and vibrated. When the concrete reached a level of 2 to $2^1/_2$ feet in the forms the vacuum was applied and maintained until half an hour after concreting had been completed. Then the forms were tightened to compensate for the compaction that had occurred during dewatering.

As to the resistance to heat and blast, it is expected that the dewatered concrete will withstand six to eight space shuttle launches before extensive repairs are necessary. Without the blast, though with the heat, your soaking pit ought to last longer.

Concrete to withstand hot metal spillage

Q. *Is there any patching material for concrete floors that could be used in areas subject to spillage of molten aluminum? The epoxies burn, the hydraulic cements blow up, and we are at a loss as to what to recommend. Can you or any of your readers give us any assistance?*

A. You should consider the use of high alumina (calcium aluminate) cement concrete. When calcium aluminate cements are mixed with suitable aggregates they produce refractory concrete that can be used at temperatures sometimes in excess of 3500° F. Such materials are often used as heat-resistant linings for power plant stacks, steel furnaces, and petrochemical heaters.

Chemical attack

Concrete pipe exposed to acid

Q. *Our pipe company has been asked to produce pipe to resist an acid solution with a pH of about 5.2. How can we calculate what the life expectancy of the pipe should be? What admixture, if any, would make the concrete resistant enough to prolong its life? Would it help to use sacrificial concrete?*

A. Calculation is an unlikely way to get an answer because there are too many unknowns of temperature, rate of flow and the rate at which reaction products are removed and new surface is exposed. We do not know of any admixture, other than a polymer used to make polymer concrete, that will significantly improve the acid resistance of portland cement concrete. Fly ash and other pozzolans will help, but will probably not give the amount of increased life needed. Switching from portland cement to calcium aluminate cement might also be helpful. Sacrificial concrete (made by using limestone as the aggregate) would prolong the life to some extent, but possibly not to the extent needed in a flowing system where there is an essentially inexhaustible supply of acid. Some pipes have survived for substantial periods in nonflowing, mildly acidic waters, as described in Hubert Woods' *Durability of Concrete Construction*, American Concrete Institute Monograph Number 4, pages 123-126.

The best way to ensure pipe longevity is to use a coating or liner. ACI 515R-79, "A Guide to the Use of Waterproofing, Dampproofing, Protective, and Decorative Barrier Systems," provides further information.

Calcium chloride likely to corrode metal deck

Q. *Through an error, 1 percent calcium chloride was added to a 3000-psi concrete mix which was used for a slab over a galvanized metal deck. The specification clearly prohibits the use of the admixture. The structural engineer and the architect have directed us to either remove the slab or show that there will be no detrimental effects from the addition of the calcium chloride. The removal would cause unacceptable delays to completion and occupancy. Is the calcium chloride detrimental to the concrete or the deck?*

A. The chloride will have no detrimental effect on the concrete itself but it is likely that the concrete would eventually corrode the metal deck. In the article "The Role of Calcium Chloride in Concrete," *Concrete Construction*, February 1976, page 57, Bernard Erlin and William G. Hime cite six case histories of corrosion from calcium chloride. In one case interior steel door jambs were inset in concrete containing 1.4 percent calcium chloride dihydrate by weight of cement. Within 1 year after construction the steel jambs were eaten through in many areas and moderately corroded in others.

Will low pH wastewater attack concrete?

Q. *We're building a wastewater treatment plant with two concrete tanks that hold wastewater having a pH between 5 and 5.5. Is concrete adequately resistant to acids of this strength? Or do we need to use a protective coating?*

A. The question isn't easy to answer simply. By itself, pH isn't a true indicator of the aggressiveness of water. The pH has no direct relationship to the amount of acid present and therefore it can't be used alone to predict the degree of attack by the water. The neutral point on the pH scale is 7.0; values below 7 indicate acidity and values above 7 alkalinity. Water with a pH in the range of 7 to 8 is alkaline, but if it can dissolve calcium hydroxide it may still be aggressive to portland cement concrete. The dissolving power of such water depends on the relationship of total dissolved solids, temperature, hardness, alkalinity, and the pH of the water (Ref. 1).

The Aggressiveness Index (AI) is a useful tool for determining whether or not a particular water is potentially damaging to concrete. The AI is defined in the American Water Works Association (AWWA) Standard for Asbestos-Cement Distribution Pipe (Ref. 2):

$$AI = pH + \log_{10}(AH)$$

where A = total alkalinity in mg/l $CaCO_3$

H = calcium hardness as mg/l $CaCO_3$

According to the AWWA document, the AI can be used to classify the water's potential effect on concrete as:

Highly aggressive	$AI < 10.0$
Moderately aggressive	$10.0 < AI < 11.9$
Nonaggressive	$AI > 12.0$

Calculating the AI requires more information than just the pH, but it can be more useful than relying on relationships between pH and concrete damage. Concrete exposed to highly aggressive water would probably require a protective coating. Moderately aggressive water could probably be handled by using good quality concrete and good construction practices

- Enough cement and a low enough water-cement ratio to produce dense concrete
- Air entrainment to provide freezing and thawing protection and reduce permeability
- Fly ash or other pozzolan that will combine with calcium hydroxide to form insoluble lime silicates
- Design for watertight contraction and construction joints that minimize intermediate cracking
- Adequate consolidation during placing
- Moist curing

More information on the effect of chemicals on concrete is found in a pamphlet by the Portland Cement Association, "Effects of Substances on Concrete and Guide to Protective Treatments" (Ref. 3).

References

1. Green, J. K., and P. H. Perkins, *Concrete Liquid Retaining Structures—Design, Specification and Construction,* Applied Science Publishers Ltd., London, England, 1979.

2. AWWA Standard for Asbestos-Cement Distribution Pipe, 4 inch through 16 inch, for Water and Other Liquids. AWWA C400-77, American Water Works Association, Denver, Colorado, 1977

3. "Effects of Substances on Concrete and Guide to Protective Treatments," IS001.04T, Portland Cement Association, Skokie, Illinois, 1981.

Deterioration from magnesium chloride

Q. *In Problem Clinic in the February 1982 issue you made recommendations for concrete to be used for bins in which rock salt (sodium chloride) was to be stored. We are having major problems with cracking, spalling and surface weathering of concrete used in salt storage. In this service the salt is being stored in damp conditions and it contains magnesium chloride. We are very interested to note that you consider the corrosive action of magnesium chloride to be more serious than the action of sodium chloride.*

It would help us with our research into the problem if you could expand your answer further, stating previous research and the reasons for your conclusions.

A. The critical statement in the earlier Problem Clinic item is "Although there is some slight possibility of expansion cracking [from absorption of sodium chloride solution], the more serious possible effect is from magnesium chloride, often found in rock salt as an impurity. This may cause cracking, followed by corrosive attack on the reinforcing steel." This statement is based on an item in Table A3 of ACI 515, "Guide for the Protection of Concrete Against Chemical Attack by Means of Coatings and Other Corrosion Resistant Materials," published in the *Journal of the American Concrete Institute,* December 1966, pages 1305-1392. The information actually came from earlier sources.

Since you are handling damp salt the concrete may in effect contact a solution saturated with sodium chloride and containing magnesium chloride in some lower concentration. ACI 201.2R-77, in a table showing the effect of commonly used chemicals on concrete, lists magnesium chloride solutions as having a slow rate of attack on concrete at ambient temperature. F.M. Lea, in *The Chemistry of Cement and Concrete* (third edition, Chemical Publishing Company, New York, page 673) says: "Solutions of magnesium chloride of 2 percent concentration and upwards produce a gradual diminution in the strength of portland cement mortars," and "Very strong [about 30 percent] solutions destroy portland cement concrete."

More general comments are given by Hubert Woods, in *Durability of Concrete Construction, (American Concrete Institute Monograph Number 4,* 1968, page 127). He says "The chlorides and nitrates of ammonium, magnesium, aluminum and iron all attack concrete, with those of ammonium being the most harmful." This is a summary of information based on the F. W. Locker and H. Pisters article, "Assessment of Waters Aggressive to Concrete," *Zement-Kalk-Gips* (Wiesbaden), Volume 17, 1964, pages 129-136.

Data showing considerable expansive effects of magnesium nitrate and magnesium sulfate solutions were reported by William H. Kuenning in "Resistance of Portland Cement Mortar to Chemical Attack-A Progress Report,"Portland Cement Association *Research Bulletin 204* (reprinted from *Highway Research Record* Number 113 [1966], pages 43-87). On the basis of several literature references the author also said "Most magnesium salts are destructive to concrete. They leach out calcium hydroxide to form the soluble calcium salt and deposit magnesium hydroxide, which is far less soluble than calcium hydroxide. They are also said to react with other cement hydration products to form such compounds as magnesium hyrdoxide, soluble calcium salts, silica, and alumina; the latter two possibly react with the magnesium hydroxide to form hydrated magnesium silicates and magnesium aluminate."

Most of the references cited here contain additional bibliographic material that may be useful to you.

Cleaning

Eliminating urine odors

Q. *What method should be used to rid concrete of the odor of urine?*

A. Where concrete is repeatedly wet by urine, frequent washing of the concrete has been advocated, but because the urine tends to penetrate the concrete this doesn't seem to be a satisfactory solution, even if strong detergent, oxidizing agents or deodorants are used.

Probably the best solution is to coat the concrete with an epoxy. This seals the concrete with a coating that is not attacked by urine. This surface is easy to wash after wetting by urine. An epoxy manufacturer could make a recommendation about which of his epoxy coatings would be most suitable.

Cleanup after a fire

Q. *A fire broke out in our warehouse but it was extinguished within an hour by the fire department. We were glad to learn from investigating engineers that the walls, which are made of tilt-up construction, are still structurally sound although they are covered with black sooty stains. Is there a good way to remove the stains so that the walls will be light in color? For most of the area a simple cleaning will be satisfactory to restore light reflectiv-*

ity. However, we will want to clean some areas near the office well enough so that they can be painted.

A. Most smoke stains can be removed readily by scrubbing with scouring powder and water. This can be followed by swabbing with a cloth or sponge soaked in a solution of sodium hypochlorite. Workmen should be careful to wear rubber gloves to protect their hands and goggles to protect eyes against splash. The sodium hypochlorite should be rinsed off, especially in the areas where paint will be applied.

Removing oil stains

Q. *What is the best way to remove motor oil stains? The oil has probably soaked pretty deeply into the concrete.*

A. Either mineral spirits or a strong water solution of a detergent (1 part liquid detergent to 5 parts water) can be used. But a good result cannot be obtained simply by applying these materials by themselves to the concrete, even with scrubbing. The problem is that they carry part of the oil even more deeply into the concrete.

The mineral spirits or the detergent solution can be used, however, to make a poultice that can be plastered onto the surface. Fly ash, hydrated lime, or other inexpensive finely divided material can be used with mineral spirits or the detergent solution to make a paste to be used as a poultice. Shredded newspaper, paper pulp or sawdust are also useful. As much liquid should be used as needed to make the poultice trowelable or moldable. The product is applied to the surface and left in place.

The liquid slowly migrates into the concrete and dissolves some oil. Meantime, as the poultice begins to dry, the dissolved or emulsified oil begins to move out of the concrete. As the solvent or water evaporates, it leaves its load of oil in the poultice. When dry, the poultice can be lifted off the concrete and discarded.

The advantage of the detergent is that it doesn't produce objectionable fumes and can be used where solvents are banned. But it takes more treatments to get the same result as with mineral spirits. *Mineral spirits must not be used without excellent ventilation and care must be taken to prevent explosions or fires caused by sparks* (for example from an electric fan).

Removing chewing gum

Q. *Please send information on how to clean chewing gum from concrete sidewalks. I have a steam cleaning machine that operates at 650 psi and a cold water cleaning machine that operates at 2000 psi. Either machine will remove dirt, grease, ice cream, dried soft drinks, and chewing gum. But both machines leave an outline on the concrete where chewing gum has been, and I have not yet found a way to remove the stain. My method has been to spray the concrete with a mixture of a multipurpose powdered cleaner and cold water. After allowing this to stand on the surface about 20 minutes I clean with the steam machine if there is a lot of gum or the cold water machine if there is not. The result is fair, but not satisfactory to me. I do not want to use muriatic acid because I am usually working in the vicinity of a lot of people, cars, trees, and plants.*

A. Manufacturers of cleaning materials suggest two different methods of removing the gum. One method is to use a freezing agent made for the purpose. When this is applied by spraying from an aerosol can, the gum freezes rapidly and can be lifted off immediately while it is brittle. It is said that a 12-ounce can will remove a lot of gum. It is not entirely certain whether the stain will be left behind, but one manufacturer believes not.

The other method is to use any of several high-powered cleaner formulations. One company suggests using its paint stripper for gum removal; this is reported to be a blend of emulsified solvents. This material is diluted with water, applied to the concrete, left in place for up to 30 minutes, and then removed with a high pressure hot or cold rinse. Another company recommends an industrial grade solvent formulation that is used to remove grease, stains, and epoxies, and which does some slight etching of the concrete. Another company has several formulations which are considered effective.

Restoring the surface of a historic structure

Q. *As the consulting architect for the restoration of a small but very important historical Hawaiian chapel structure (built in 1864), I have given considerable thought and research to the problem of cleaning the original cement plaster exterior finish of the many years of algae growth and its resultant discoloration (dark gray to black). This cleaning must be done without defacing or altering the surface sand texture of the plaster. For example, sand blasting and acid application would be unacceptable. The most rigid tool permissible would be a wire brush. In addition to this cleaning, we have the task of preserving the plaster itself. While small patches that are decayed and crumbling will be replaced, over 95 percent appears to be in preservable condition. If possible we would like information on a suitable cleaning agent and a long-lasting protective coating.*

A. There are now a considerable number of manufacturers of cleaning compounds for concrete. The Buyers' Guide in the December issue of *Concrete Construction* lists names and addresses of manufacturers. Probably not all of these companies manufacture a product suitable for your purpose but certainly some of them do.

Regarding patching, you may want to give attention to methods of color matching. This is usually accomplished by making a number of standard mixes which all have the same ratio of cement to sand but different ratios of gray to white cement. These serve as a guide to what is used on the job. A useful set of mixes could contain 1 part gray to 0 part white cement, and 3 to 1, 2 to 1, 1 to 1, and 1 to 2 respectively. These should be given the same float or trowel finish that will be used on the patches. It is likely that you will want to change the patching mix from one location to another to make the best match with surrounding concrete. If you plan to use acrylic latex or similar material in your patches you would of course use it in the standard mixes.

Albert Litvin, of the Portland Cement Association, ran some tests a number of years ago that showed that methyl methacrylate coatings tend to be the best coatings for such applications as yours, though not necessarily every methyl methacrylate formulation. We suggest you ask the manufacturer for evidence that the coating you choose does not yellow or crack with age. Litvin's results were reported in the article, "Clear Coatings for Exposed Architectural Concrete," *Concrete Construction*, May 1969, page 169.

Art of restoring old buildings

A reader, Gustave Rener, II of Gustave Rener and Sons, Grosse Pointe Woods, Michigan, has written to comment on the Problem Clinic item about the Hawaiian chapel, page 352 of the April 1983 issue, "Restoring the Surface of a Historic Structure." His response is given below.

When cleaning historic cement plaster to prevent damage, always start with the most mild method of cleaning and proceed to stronger methods. Use small test patches until a satisfactory surface is restored.

Always start with:

1. water with soft brushes
2. water with a mild soap
3. water with stronger soap
4. water with stronger soap plus ammonia
5. water with stronger soap plus vinegar

Only after trying these methods in this sequence should you proceed to the stronger chemical cleaners.

Modern cements themselves are often radically different from the cements used historically. The early cement used in the Hawaiian chapel could possible have been homemade from oyster shells or it could be a natural portland cement. Investigation should be done to determine its characteristics, and then materials available now should be color blended to formulate a material resembling the present state of the old plaster.

If cement color matching is carefully done the predominant factor remaining is the aggregate. In color matching mortar, the hardest and most important thing is the selection of aggregate. A small sample should be taken and checked to ensure as close a match as possible. Coloration, grain size, and grain shape will all affect the finish, color, and texture as much as the cement itself, if not more. Unless color matching is done, what looks good in a sample today will alter significantly after a few years.

Modified mortars must be given a bit of thought before using. Quite often their strength is much higher than the original plaster. This results in spalling where the patches meet the original surface. You must match similar strength materials to similar strength material to achieve a sound, long-lasting patch.

More on algae and old buildings

Regarding the Hawaiian chapel discussed in the Problem Clinic item on page 352 of the April 1983 issue, and again in September 1983, page 712, further correspondence reveals that the structure was built in 1864 and the cement plaster finish is in fragile condition. The substrate is coral block masonry which is in good condition. Spots of cement plaster have deteriorated and spalled here and there. The coral blocks were set in what is believed to be homemade mortar of ground coral (because it has a high gypsum content) baked in a charcoal fire. At that time cement could be obtained only from the east coast of North America via a long sea voyage around the southern tip of South America. The block was probably left exposed until about 1900, when the cement plaster finish was applied.

Over the years the cement plaster has become badly discolored with black streaks of algae. Bleach destroys the algae but does not seem to have any significant effect on the stain.

Thomas C. Geary, Consultant, of Redwood City, California has sent extensive comments that include the following:

I suggest you check for additional looseness of plaster over coral blocks in areas that now appear to be in good condition. In a very old building, bond may have been lost in areas where it is not immediately obvious. You can check by rapping gently on the walls with a length of narrow steel rod, a steel chisel, or a piece of hardwood rod. Rap in many locations. A hollow sound will be apparent when you tap over areas of looseness, but a hard, rock-like sound will emanate where bond to the substrate is good. Any areas suspected of loss of bond should be removed.

Mildew grows on concrete and thrives in shaded areas. Although you say that bleach (sodium hypochlorite) does not affect the stain, when you employ bleach, I suggest a mixture of about 3/4 cup of bleach in 1 gallon of warm water. Do not mix any other cleanser or agent with bleach. After treating with bleach, allow perhaps 15 to 30 minutes exposure to affect the spores of algae, then rinse the bleach away carefully. Finally, scrub the concrete vigorously with a nylon brush and a strong detergent. I do not recommend a fiber brush because it gradually loses stiffness when exposed to warm water for a prolonged period. Do not allow anyone to persuade you that a steel wire brush should be used for this purpose. Small particles of steel will abrade and be deposited upon the plaster to form brownish rust stains later.

It may be that vigorous scrubbing will remove killed spores of mold. If not, light blasting with particles of nut shells in sandblasting equipment will remove the stains. If you must resort to blasting, the nozzle should be held further than normal from the surface of the plaster to reduce abrasion. Water blasting is also sometimes effective.

If none of these methods completely removes mildew, an overlay of cement plaster should obscure the stains. All loose algae must first be removed in order to get good bond to the substrate.

Restoration of old buildings

Various aspects of restoring the surface of a historic Hawaiian chapel have been discussed in the issues of April 1983 (page 352), September 1983 (page 712) and January 1984 (page 59). Gustave Rener, who contributed to the September 1983 discussion, now adds the following:

We must understand the difference between restoring concrete or stucco structures and the *historic restoration* of these structures. Techniques, materials, and workmanship that are perfectly acceptable in the restoration of a normal structure such as a parking garage or building are sometimes unacceptable in historic restoration. The principal difference is that historic restoration requires returning the fabric of the structure to what it looked like originally, or at its most important historic date.

Modern materials are perfectly acceptable for patching individual areas, but their texture and color must match the existing texture and color of the surrounding areas.

This is the big problem. The best way is like-to-like patching, repairing the structure with materials as nearly the same as the original as possible.

Because of a loss of surface character and color, sandblasting is not widely used in stucco restoration work. No matter how professionally the work is done, the fluctuations in pressure, angle of attack, degree of hardness, and size of shot used will produce an uneven surface. The hardness of the original stucco itself varies because in its historical context it was probably mixed and applied by craftsmen who were less than properly trained and supervised. The sandblasting could tear out pockets of soft material that is structurally sound but less dense than the surrounding material.

I would suggest that if cleaning the surface will not remove all the dirt and stains, leave it alone. Remember: clean first, then patch. In this way the patching can be closely matched to the finished look. Destruction of the fabric of the building while restoring is one of the principal reasons for withholding the designation of a building as a historic structure, and for the resulting loss of tax credits.

Removing roofing tar

Q. *A considerable area of our parking lot has been messed up with roofing tar—some of it windblown from the roofing operations and some of it dribbled from buckets and kettles. Cleaning with dry ice and scrapers would probably be slow and tedious. Is there a faster way?*

A. Investigate water blasting. Manufacturers of water blast equipment can tell you whether it will produce a satisfactory appearance, how fast it will work, and perhaps give tips on technique.

Removing oil stains from pavements

Q. *The stalls in our company parking lot have become unsightly because of oil stains that have accumulated over many years. Is there an effective way to remove the stains?*

A. One method has been described in Problem Clinic in the April 1982 issue, page 386. More recently, similar methods have been studied by the National Ready Mixed Concrete Association and these have been described in the January 1984 issue of the Association's *Promotion Pointers.*

The study investigated the use of three agents:

- a solution of 5 percent sodium hydroxide to be used with ground limestone
- granular trisodium phosphate crystals to which hot water was added
- a proprietary driveway cleaner and degreaser containing sodium metasilicate and petroleum distillate

Tests were conducted on a pavement where oil spills had been accumulating for 20 years. A heavily stained area was selected and scraped with a putty knife to remove accumulated grease and dirt. The area was then marked off into three parts so that one agent could be tried on each part and compared with the other agents.

The sodium hydroxide solution used can be made from sodium hydroxide pellets or flakes at the rate of 0.5 pound per 10 pounds of water (or 6.5 ounces per gallon). The solution was mixed with enough ground limestone to make a stiff paste. This paste was then spread over the stain with a putty knife and allowed to dry completely. It was subsequently brushed off.

The trisodium phosphate crystals were sprinkled on the stain until the stain was completely hidden. Hot water was added and the area was scrubbed with a stiff brush. After 30 minutes the area was thoroughly rinsed.

The commercial driveway cleaner and degreaser was scrubbed into the stained area and then the area was thoroughly rinsed.

The results were all adjudged to be good. The paste of sodium hydroxide and ground limestone was the most effective but was also the most time-consuming, both in the time required to apply it and in the time needed for the drying process, which takes at least 24 hours. The trisodium phosphate and the driveway cleaner have the advantages of being easy to obtain, and fairly easy to apply, yet productive of good results.

Sodium hydroxide can be purchased from chemical supply houses or, in the form of lye (stocked for cleaning household drains), from a hardware store. Ground limestone can be obtained from various sources such as materials yards, limestone quarries, garden and lawn supply centers, or hardware stores. Hardware stores are also the source of both trisodium phosphate and the particular commercial cleaner that was used (which contains sodium metasilicate and petroleum distillate).

Modern concrete to replace yesterday's wood

Q. *We have had an increasing number of requests to replace factory woodblock floors with concrete floors. There might actually be enough of this kind of business to keep us occupied full time. One thing that makes us hesitate is that we aren't sure about whether there is a good way to clean up the base slab effectively enough to bond the new concrete topping to the concrete base slab. What can you tell us?*

A. This subject has recently been getting attention in the Indiana area. The June/July 1984 issue of the Indiana Concrete Council's *Concrete Columns* reports that in Marion, Indiana a contractor investigated this problem. He learned from the manufacturer of the wood blocks that a scarifying machine is effective for removing the mastic that bonds the blocks to the floor and cleaning the base slab. The old, brittle mastic is said to come off easily, although they say that newer mastic should be blotted with silica sand before trying to remove it this way. Both walking model and riding model scarifiers have been used successfully. Machines with widths of cut up to about 20 inches can be obtained from local rental yards.

The Indiana Concrete Council also suggests that blasting with water might be effective and they refer to the June 1984 issue of the Portland Cement Association *Concrete Technology Today*, which reports on removing old floor tile adhesive but not specifically on removing the mastic used to hold woodblock flooring. That article says that a water blaster with a pressure pump that provides pressure from 800 to 2000 psi produces excellent results. Using hot water at about 180° F improves the efficiency. Then the water and refuse are picked up with a wet vacuum system.

Cleaning sealant off white concrete

Q. *We constructed some white flatwork with 3/4-inch redwood separators between panels. The top 3/4 inch above the redwood was caulked. This white concrete adjoins an area of pattern-stamped dark gray concrete, so we were asked to use a black caulking material above the redwood on that side. The caulking material was self-leveling and in two places it ran downhill out of the joint onto the white concrete. We haven't been able to remove it with solvents. What do you suggest?*

A. Probably the most effective way would be to try to burn it off with an acetylene torch with a wing tip that produces a broad flame. Don't let the flame stay too long in one place and damage the concrete. Try to develop a light touch that is yet heavy enough to burn off the caulk. There may be some residual ash that may have to be brushed off with a stiff brush or broom.

If that proves unsatisfactory, we suggest asking the manufacturers of the caulking material for suggestions on how to remove their product and also asking the same question of various manufacturers and cleaners (see "Cleaning Materials" in the Buyers' Guide of our most recent December issue of *Concrete Construction* magazine). However, if you do find an effective solvent it will probably be necessary to use it in combination with white rags or cotton, weighted down and left on the concrete to absorb the caulk as the solvent evaporates. Otherwise the solvent might just carry any dissolved material more deeply into the concrete. It may take more than one application for the process to be effective.

Saw blade speed

Q. *When sawing concrete with a 30-inch diamond blade the cutting has been very slow. We've been told that we should reduce the motor speed for large blades to make them cut efficiently. Is there a rule of thumb on how fast to run a saw?*

A. The Concrete Sawing and Drilling Association gives the following suggested spindle speeds for various sizes of blades:

Blade diameter, inches	Spindle speed, rpm
12	2547
14	2183
18	1698
20	1528
24	1273
30	1019

These speeds are usually printed on the blades but sometimes wear off. They are the appropriate speeds for a blade tensioned for 9600 feet per minute total travel of a point on the circumference. They are somewhere in the range of 9000 to 11,000 feet per minute. For larger blades the speed can be calculated from the formula:

$$\text{Speed} = \frac{12 \times \text{feet per minute}}{3.14 \times \text{diameter in inches}}$$

When cutting more than 20 inches deep, the Concrete Sawing and Drilling Association suggests that the contractor install an intermediate shaft between the spindle and the engine. In general the speed should be increased for soft, abrasive material but decreased for hard, dense material. Further information can be obtained from the blade manufacturer.

Effect of grinding on durability

Q. *We're having our driveway ground because of the large number of birdbaths produced in the surface during construction. I'm a little worried about what this will do to the durability. I'm pretty sure the air content is as high as it ought to be, but how will the ground pieces of aggregate survive freezing and thawing? Our northern winters are fierce.*

A. A report by William Perenchio, "Effect of Surface Grinding and Joint Sawing on the Durability of Paving Concrete," published in the *Journal of the PCA Research and Development Laboratories*, January 1964, pages 16 to 19, says that durability was not affected when thin layers were ground off. If the grinding goes deep and exposes large areas of the coarse aggregate it may affect the durability. If this describes your driveway you may wish to treat it with linseed oil to improve the durability. Dilute the first coat with an equal amount of mineral spirits and mop it on but don't leave puddles. Dilute the second coat somewhat less. It won't hurt to apply additional coats every year or so. The concrete will be darker in color but should last longer.

Cutting a mass concrete wall

Q. *Would it be efficient to use concrete coring drills to cut a 3½x8-foot opening in mass concrete? The wall is 6 feet thick at this point, so sawing would not be possible.*

A. Yes. ASC Corporation, a general contractor in Boise, Idaho recently used two 7½-horsepower diamond core drill machines to cut a 50x90-inch opening in an 84-inch thickness of a dam wall at the Pelton Dam in Warm Springs, Oregon. The opening was 70 feet above ground, but there was a 20x20-foot pedestal just below it that served as a working platform. On this job the concrete had to be cut and removed in two pieces to limit the weights to about 8 tons each, small enough to be handled by a 325-ton crane when its boom was extended 280 feet.

A critical part of this job was the safe removal of the two large pieces of concrete. This was accomplished by inserting into the core-drilled holes two pieces of greased $5^1/_2$-inch extra-heavy pipe to serve as tubular tracks on which to slide the top piece out. It was then removed readily with cables and two 400-Hertz electric winches. The lower piece was cut and removed the same way. The tubular tracks may be seen in temporary storage above the lower piece of concrete in the photo.

The job required 64 cuts 6 inches in diameter and 84 inches deep, a total of nearly 450 feet of drilling. The time required to make all cuts and remove the concrete was reported to be 80 man-hours.

A 50x90-inch opening is cut in the 7-foot-thick dam wall.

When to grind a defective surface

Q. *Will it hurt to grind concrete the day after it's placed? We're the general contractor for an office building project. We told the flatwork contractor to use a broom finish for a wide sidewalk around the building perimeter. He misunderstood and gave it a rake finish. We want to try to save the slab, if possible, by grinding.*

A. We posed the question to Leo Swan of Equipment Development Company, Inc. Leo says you can usually start grinding within 8 hours of placement. Watch for tearing of aggregate particles and stop grinding if this happens. A rake finish probably will have furrow depths of $^3/_8$ inch or so; grooves in a broom finish generally don't exceed $^1/_{32}$ inch. You may be able to grind just enough of the rake finish away to leave shallow grooves that will still give needed traction. However, aggregate will be exposed. If the resulting surface isn't acceptable to the owner, you could try texturing it with a drum-type planer with small carbide teeth.

Saving cows

Q. *I lost an expensive cow when she slipped on manure on the dairy barn floor and couldn't get up. Is there a sure-fire way to make the floor safe for cows?*

A. We can't guarantee against all eventualities, but on a floor that is reasonably well maintained it works to cut grooves in the alleys running in the same direction as the traffic. One company recommends grooves $^{1}/_{2}$ inch wide and $^{3}/_{4}$ inch deep spaced $3^{3}/_{4}$ inches on centers. Another recommends grooves $^{5}/_{8}$ to $^{3}/_{4}$ inch wide, $1^{3}/_{4}$ to $2^{1}/_{4}$ inches on centers. You can get planing machines at many rental yards that are equipped to be set to cut grooves of these sizes and spacings. Some machines cut grooves in a 2-foot swath at the rate of about 5 feet per minute. Experience indicates that the cutting can even be done while the cows are in the barn; the noise doesn't seem to bother them. However, you ought to use a catalytic exhaust attachment to keep from polluting the barn atmosphere with engine fumes.

Slipproofing a surface for cattle

Q. *A dairy farmer has installed some steel-troweled 20- by 90-foot runners in three stalls of his cattle barn. The cows are slipping on the accumulation of manure and the problem is becoming serious. What is the best way for us to roughen the concrete for him?*

A. Probably the best solution is to scarify the surface. Several concrete scarifying machines are available commercially. An alternative is to coat the concrete surface with epoxy and scatter emery over the epoxy before it hardens. This alternative seems likely to be more expensive and perhaps less effective than the first.

Slipproofing for cattle

This response to an answer given in the March Problem Clinic, page 294, is from Rino Sclippa, Marketing Manager for the Concrete Coring Company.

One major alternative was overlooked in the response to the dairyman who had a problem with cows slipping on concrete made slick by accumulation of manure and moisture.

Concrete becomes slick and worn with years of service. The most practical way to eliminate this hazardous condition is to safety-groove the area. Over a period of many years safety-grooving has proven its value in both highway and commercial use by reducing the likelihood of accidents on wet concrete or asphalt surfaces.

Years of research and testing have gone into perfecting the specifications for precise spacing and depth of grooving to give maximum traction for foot, automobile, and livestock traffic. Because safety grooving has proved to practical and economical in preventing moisture-related accidents, it has been adopted by most state and federal agencies. It has several primary advantages. It can be orietated for drainage, is easier to clean, and stays cleaner. Moreover, its longevity makes it more economical than other methods in the long run.

The Concrete Sawing and Drilling Association, 6077 Roswell Road, Suite 205, Atlanta, Georgia 30328, will provide specifications and standards sheets covering this subject.

Followup on slipproofing for cattle

Rino A. Sclippa had an update to his letter about slipproofing.

Since the time when you published my letter in the June issue, page 537, Dennis Neff, executive secretary of the Concrete Sawing and Drilling Association, and Peter Silveri, executive manager of the International Grooving and Grinding Association (IGGA), have pointed out that there is no one simple, universal grooving standard to use for hoof traffic. I find that Lester F. Kuzmick, president of Cushion Cut Inc., Torrance, California, presented a paper on pavement grooving, grinding, and texturing at a 1981 World of Concrete seminar. He said, "There is no standard pattern such as in highway grooving, since the specific requirements vary from job to job. There are, however, some commonly used grooving patterns. The grooving and grinding specialist is best suited to recommend the pattern most suited to a particular application."

The commonly used grooving patterns follow these basic specifications: for large-hoof animals like cattle, a groove 1/2 inch wide, 3/8 inch deep, spaced 3 3/4 inches center to center; for small-hoof animals like hogs and sheep, a groove 1/4 inch wide, 3/8 inch deep, spaced 2 inches center to center. These patterns are typical of some in use. Whether they are sawed longitudinally, transversely, or both, depends upon the job conditions.

Positive waterproofing for roof

Q. *In concrete buildings I am designing it is important that exposed concrete decks over interior spaces be free of leaks. I need a tested and proven admixture or sealer to provide a permanently waterproof deck. The deck will be exposed concrete, subject to traffic and to the Florida sun.*

A. There is considerable difference of opinion about whether admixtures can effectively reduce the permeability of high quality concrete. Certainly the watertightness of any concrete, with or without an admixture, is lost if the concrete should crack. For this reason a sealant is usually chosen where leakage cannot be tolerated.

Methods of sealing concrete against water are discussed in the articles "Waterproofing Materials for the Earth-Sheltered Contractor" and "Construction Recommendations for Moisture Control," both of which appeared in the September 1980 issue. These supply information that should be useful for your purpose.

You might want to consider post-tensioning, which should guarantee a crack-free roof if done properly. (See the article "Post-Tensioning: It Makes Sense for Earth-Sheltered Homes," February 1981 issue, page 105.)

Stopping seepage through chimney

Q. *I've been told that your issue of last October had an article on waterproofing, but I don't know if it had anything to do with bricks. I am having difficulty with water seeping through my chimney. I would appreciate any information you could send me about coatings that would stop moisture seepage through bricks.*

A. We have little information on brick masonry since we concentrate on cast-in-place concrete and associated subjects. In April 1981 we published a series of three articles on the general subject of waterproofing concrete which you might find informative:

"Waterproofing: Who Needs It?" pages 305, 307

"Dampproofing: Why, Where and How?" pages 309, 311

"Waterproofing: Establishing a Barrier," pages 313, 315, 317-319

"Waterproofing Checklist," page 321

However, waterproofing a brick chimney may not be a simple matter of choosing and applying the right coating. Although untreated bricks do absorb water and although water does move through untreated bricks, it is probably much more common for seepage to come through cracks between the mortar and the bricks. If this is what is happening in your chimney, perhaps the problem can be remedied by tuckpointing. Another common cause of water problems in chimneys is cracking of the chimney cap or a separation between the chimney cap and the flue. Rainwater penetrating a cracked cap may later appear on the chimney surface at a considerably lower level. The same tradesman who does tuckpointing repairs should be able to repair the chimney cap.

If it turns out that you do need to treat the whole chimney surface you might want to use a clear, breathable coating. Or you might choose a water repellent such as a silicone or a stearate. These will reduce permeability of the brick to liquid water, yet allow water vapor to escape. They will not seal cracks between brick and mortar.

Repair of leaky dome

Q. *For the past 5 years our 12-year-old cathedral dome has been showing moist spots on the interior surface whenever it rains or snows. The moisture is not uniformly distributed over the surface, so it is not condensation. About 3 or 4 years ago a thin seal coat was applied to the exterior surface but it didn't reduce the leakage. What do you recommend for repairing and sealing the dome effectively?*

A. We suggest you ask two expert waterproofing contractors in your area to examine the dome and make recommendations as to whether or not you have small cracks that need to be repaired by injection. If there are small cracks that are of no structural significance the contractors may be able to seal them by injecting a urethane foam designed for the purpose. Structural cracks may have to be repaired by injecting with epoxy. If the problem is not too severe it may be possible to prevent leakage without injecting the cracks but by applying a neoprene-hypalon coating system to the exterior.

Cause for peeling of urethane coating

Q. *We poured a concrete floor slab for a 40,000-square-foot tilt-up building. About 5 months later, a subcontractor applied a urethane coating that soon started peeling and deteriorating. Now portions of the coating are being removed. A shotblaster was used first but it started taking out the concrete surface too. They switched to a scabbler with carbide heads but it still removed concrete as well as coating. The subcontractor says there's a weakness in the concrete and that's why the coating failed. He says either the concrete was too wet or it was overfinished. In other words, he says the failure is my fault. I don't think we caused the problem. The exposed slab went through a winter without scaling or spalling before the urethane was applied. There is still no spalling at the joints. The crane did no damage to the floor when it was picking up tilt-up panels. And there were no problems with the panels placed by the same crew that did the floor. I think the subcontractor didn't properly rinse off the acid wash he used to prepare the floor for coating. How can we find out why the coating failed?*

A. Take samples of the concrete, preferably some with part of the urethane still in place. Send them to a petrographer. He or she should be able to tell the depth of acid etching by microscopically examining the surface. The chloride content of the concrete may indicate the acid used. Other reasons for coating failure may be found: water pressure, reaction with alkalies in the concrete, or a weak surface layer. For advice as to location and number of samples needed, talk to the petrographer before taking samples.

For a listing of firms providing petrographic services check the January 1988 issue of *Concrete Construction*, page 56.

Hardening agents

Q. *We have been asked to comment on specifications that call for applying a fluosilicate hardener to concrete slabs at the time of construction. Some slabs are indoor and some outdoor, all part of an agricultural facility. Does such a hardener serve to both harden and cure? Does it act as a sealer? The specifications call for applying a membrane curing compound.*

A. ACI Committee 302 considers liquid floor hardeners (even including fluosilicate hardeners, which are probably the best) to be only remedial agents. Fluosilicate hardeners function by reacting with cement compounds to deposit crystalline fluorides and silica in the pores of the concrete. This makes the concrete harder, stronger and less porous than it had previously been. Fluosilicate hardeners are intended for use to improve the hardness of floors that are defective, such as floors that are too soft because the concrete mix was too low in strength design, floors that were not placed, finished, or cured properly, or floors with surfaces that have become carbonated from use of unvented heaters. This type of hardener does not act as either a curing agent or sealer, and when it is applied it should not be used until after any liquid membrane curing agent has dissipated or been removed.

Fluosilicate hardener should really not be used as part of the planned construction of new floors. Instead, the concrete should be designed to produce enough strength to provide the hardness required. Good construction practices and good curing practices should be followed (see ACI 302, "Guide to Concrete Floor and Slab Construction"). A very high quality floor will have a low enough porosity so that it would not soak up much fluosilicate hardener anyway.

Nongloss sealer to repel oil?

Q. *I have just cast a concrete table using white cement, sand, and a superplasticizer. I am looking for a sealer that will protect the smooth concrete surface from oil stains without substantially altering its color or matte finish. So far, all I have found are products that darken the color or give the concrete a sheen or gloss. Do you have any suggestions?*

A. That's a tough one. Water repellents like silicones or stearates do not repel oil. The penetrating sealers we know of would darken the concrete. There are a number of clear coatings but all of them impart gloss. Among these glossy coatings are the acrylics, which have the additional disadvantage that they might soften from oil. Clear urethanes might gradually turn yellow, and clear chlorinated rubber might yellow to some extent. Yellowing would perhaps be less if the table is used indoors and not outdoors. If gloss and some likelihood of yellowing are less objectionable than oil staining, you might apply a clear chlorinated rubber. Otherwise you might have to settle for a tablecloth.

Protection against ammonium salts

Q. *We have to recommend a suitable coating for concrete tanks. Some will hold solid ammonium nitrate and some solid ammonium sulfate. What will happen if the coating breaks? Will these salts affect the concrete? If the salts do affect the concrete would the effect be diminished or eliminated if the salts were kept dry?*

A. Ammonium nitrate and ammonium sulfate, though they are salts, both behave as acids when moist. Therefore either of these salts in the moist condition would erode concrete. In addition, sulfates react with concrete to cause disruptive expansion, so ammonium sulfate attacks concrete both by erosion and disruptive expansion. If the salts were completely dry their effects would be minimized. However, most concrete, even though dry on the surface, contains a considerable amount of moisture. This moisture could evaporate from the concrete, condense in the salt, and produce enough salt solution to attack the concrete.

What is needed is a coating that, in addition to being immune to these salts, will span any crack that forms after the coating has been applied. Perhaps it should have good abrasion resistance as well. Since you are the coatings expert, and are familiar with the expected service conditions, we leave the specific choice to you.

Water storage facilities

Q. *I would appreciate any design and construction literature you have pertaining to concrete water storage facilities. I am particularly interested in specifications for watertightness and information on making joints watertight. Does concrete impart any taste to the water, and if so, what treatment can be used to overcome the taste?*

A. Concrete is commonly used without coatings for reservoirs and cisterns. Although some minerals such as calcium, sodium, and potassium hydroxide may be slowly leached out of the concrete, these appear in such tiny amounts in the water that they are likely to have less effect on the taste than the minerals the water already contains.

Our April 1981 issue contained an article, "Waterproofing: Who Needs It?" beginning on page 305. It emphasized the desirability of using concrete with a low water-cement ratio. This can be achieved by using a high cement content or a low water content or both. Other articles in the same issue discuss waterproof coatings. Although you would not necessarily need a coating, an interior coating capable of bridging any cracks that form in the future might be worthwhile.

The American Concrete Institute has a good publication, "Guide to the Use of Waterproofing, Dampproofing, Protective and Decorative Barrier Systems for Concrete," ACI 515R-79. It is a 44-page brochure available from the American Concrete Institute, Box 19150, Detroit, Michigan 48219. The same Guide also appears in the ACI *Manual of Concrete Practice*, Part 5, 1981 edition, and in the November 1979 issue of *Concrete International*.

Peeling coating

Q. *Stucco panels on one of our structures were treated with a silicone water repellent, flooded onto the surface by brush. Then an asbestos-containing coating was applied. After about 2 years the coating started peeling off in large pieces from about 6 square inches to about 1 foot square. We enclose some samples of the peelings. The architect insists that the stucco finish was not installed properly because some of it is adhering to the coating that peeled off. We believe that the stucco was painted too soon and that when the moisture could not penetrate the coating, it caused the coating to peel. What can you tell us?*

A. We do not have facilities to analyze or examine the peeled coating. We can say that peeling usually results where moisture has built up below the coating and has no opportunity to escape. If the coating does not have enough breathability for the conditions, the water vapor pressure will push it off. We suspect that is what happened on this job. The fact that a thin layer of stucco was pulled off with the paint probably means only that the strength of bond between the coating and the stucco was higher than the tensile strength of the stucco. The motive force was most likely an excessive buildup of vapor pressure below the coating. Exactly what caused the vapor pressure to build up so high is something we cannot say; it could have been too early an application of the coating or it could have been high relative humidity within the building.

Floor traffic stripes

Q. *Please send us information about a method of applying a paint to an industrial floor exposed to oil spills and heavy forklift traffic. We apply a 6-inch-wide band to indicate aisle limits and no-admission areas. Our present procedure of using a name-brand paint requires frequent maintenance.*

A. In general, epoxy coatings give good service on floors where wear resistance is important, and they are resistant to oil. You may wish to consider an epoxy system onto which fine emery is sprinkled before it sets in order to make it slip-resistant.

If the floor contains much oil already, you may have trouble getting a good bond of the coating to the concrete. The area to be coated should be scrubbed one or more times with 10 percent caustic soda solution, or trisodium phosphate solution or strong detergent, then rinsed thoroughly.

Bond may be improved by acid etching the area to be painted. This should be done *after* the caustic soda or other treatment has been completed. A 10 percent solution of muriatic acid should be applied to the area, taking care to protect the adjacent area. After the bubbling ceases, the spent acid should be thoroughly rinsed away.

In choosing an epoxy it would be best to consult with an experienced manufacturer for recommendations about the kind of epoxy most suitable for the purpose. The company may have a broad line of epoxies and the coating should be chosen with this particular application in mind.

Shiny residential floor

Q. *We are under contract to build a large custom home. The buyers want to have concrete floors kept uncovered and with a shiny finish. We have never done this before and are seeking information for varnishing the slab.*

A. You should first be sure you produce a good, smooth, hard troweled finish. You will also want to take precautions against cracking by using good jointing procedures. An adequate amount of welded wire fabric used in the slab will limit crack width in case shrinkage-temperature cracks do occur. The fabric should be positioned about 2 inches below the surface. You can find good instruction on both finishing and jointing in ACI 302, "Guide for Concrete Floor and Slab Construction," available from the American Concrete Institute, Box 19150, Detroit, Michigan 48219. A good publication on the use of welded wire fabric is *Using Welded Wire Fabric in Residential and Light Construction*, available from Wire Reinforcement Institute, 8361 A Greensboro Drive, McLean, Virginia 22102.

The coating should be an epoxy or a polyurethane coating, either of which will provide excellent wear resistance, though you should understand that not all epoxies have the same wear resistance, nor do all polyurethanes have the same wear resistance. You will have to ask the manufacturers which of their products are epoxies or polyurethanes and suitable for use as floor coatings.

Another possibility is to produce a maximum amount of shine by a sufficient number of hard trowelings, and then improve the shine by applying a slip-resistant wax.

Testing

8

Concrete

Tests on materials for concrete

What do compressive strength results on cement mill test reports tell you?

Q. *A mill test report from a manufacturer of Type I cement gives compressive strength at 3 and 7 days. Can this information be used to determine how much cement is needed for a cubic yard of 3000-psi concrete?*

A. No. The tests are made on 2-inch cubes of a standard mortar made with one part of cement to 2.75 parts of graded standard sand. The mortar water-cement ratio is 0.485. To meet ASTM's Standard Specification for Portland Cement (ASTM C 150), standard cubes made with Type I cement must have minimum 3-day and 7-day strengths of 1800 psi and 2800 psi respectively. Strengths shown on the mill test report should equal or exceed these values. But knowing the reported cube strength won't tell you how much cement is needed to produce a given strength concrete. The amount of cement needed depends primarily on the required slump and water-cement ratio, not on the cement properties.

Determining moisture in aggregate for mix proportioning

Q. *Is there an accurate, quick and inexpensive way to determine the percent of water in aggregate? I batch up to 100 cubic yards of concrete a week for our in-house projects, using all applicable admixtures. Because the water in the aggregate varies, I have to guess at the amount of water for each mix. This takes unnecessary time and extra trial batches.*

A. Here are three possible approaches:

Rule-of-thumb. Assuming your sand and gravel or stone aren't unusual in nature, you can estimate the free moisture that the aggregates may contribute to the mixing water. Sand can be expected to contribute about 2 to 6 percent free moisture by weight, and gravel or stone about 1 or 2 percent. Under average conditions that means that 104 pounds of sand would give you the equivalent of 100 pounds of sand in what is called a saturated surface dry (SSD) condition and 4 pounds (1/2 gallon) of free moisture. This 4 pounds of moisture or water becomes part of the mix water. Newly delivered sand, in the early part of the day, will have more free moisture. After a rain the sand may briefly have even more than 6 percent moisture.

The coarse aggregate contains less free moisture and varies less. You may have as little as 1 pound of free moisture per 100 pounds of gravel or stone, and seldom more than 2 pounds (1/4 gallon).

Oven drying. For quick quality control some operators have found a microwave oven to be useful. You place several pounds of the sand, gravel or stone in a Pyrex dish of convenient size, such as 9 x 13 inches. Weigh the filled dish, then deduct the weight of the dish. Place the dish in the oven for about 30 to 40 minutes at full power setting. Using oven mitts, remove the dish, cool it to room temperature (about 45 minutes) and weigh. The difference in weight represents the total moisture content of the aggregate.

For making batch calculations it is handiest if moisture has been calculated as percent of the SSD weight of the aggregate. The total moisture measured in the test represents both the free moisture and the moisture that the aggregate absorbs. Only the free moisture becomes part of the mix water. The absorbed water, usually about 1 percent, remains within the aggregate. If the coarse aggregate is unusually porous, the absorption should be determined by a separate test.

Example for test on sand:

Weight of wet sand sample . . . . 4.16 pounds

Oven dried weight . . . . 3.96 pounds

Saturated surface dry (SSD) weight,
(1% absorption)
3.96 x 1.01 . . . . 4.00 pounds

Free moisture, 4.16 - 4.00 . . . . 0.16 pound

Free moisture, percent of SSD weight
(0.16 ÷ 4.00) x 100 . . . . 4.0 percent

In the example, 104.0 pounds of aggregate would represent 100 pounds of SSD aggregate and 4.0 pounds of mix water.

A hot plate may also be used for this kind of testing but without the same fringe benefit of lunch preparation that the microwave offers.

The Speedy Moisture Tester. One of the simplest methods of testing, using equipment specifically designed for moisture tests, is the Speedy Moisture Tester, made by a British manufacturer and marketed by several concrete testing equipment suppliers in this country. The equipment is packaged so that it can be used in either the field or the laboratory. The test uses calcium carbide as a reagent and requires about 3 minutes for testing a 200-gram sample. Several tests are recommended to verify the results. The equipment is calibrated to indicate percent moisture based on the *wet* weight of the test sample within a claimed accuracy of 0.5 percent. You could use a conversion factor to change the percent moisture from the wet basis to the SSD basis.

You may be interested in Chapter 3, "Properties of Freshly Mixed Concrete as Influenced by Aggregate," in ACI 221R-84, *Guide for Normal Weight Aggregates in Concrete,* published by the American Concrete Institute, Box 19150, Detroit, Michigan 48219.

Tests on hardened concrete

Load test a questionable slab

Q. *A reinforced concrete slab consisting of very narrow tee beams was poured in place with hardly any quality control at a very low slump and no vibration. This resulted in a porous finish and even cavities within the web. Should this be a matter of structural concern? (See the cross-sectional sketch of the slab and of the entire structure.) How can these problems be corrected?*

A. We believe this should be a matter of structural concern. Whenever concrete is placed without quality control, without vibration, and with void formers that conceal the surface of the concrete, there is reason for concern about its structural integrity.

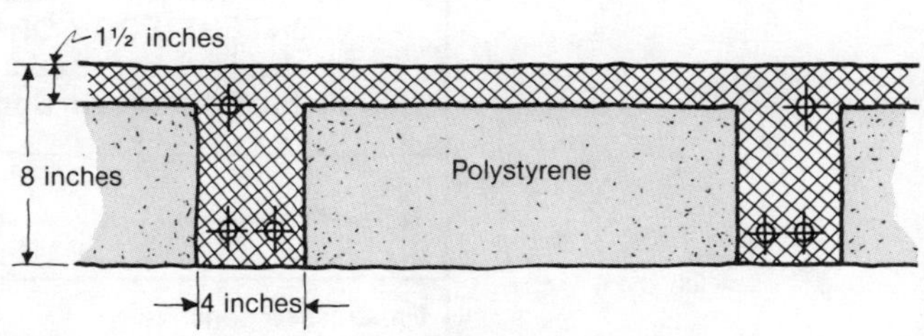

CROSS SECTION OF SLAB

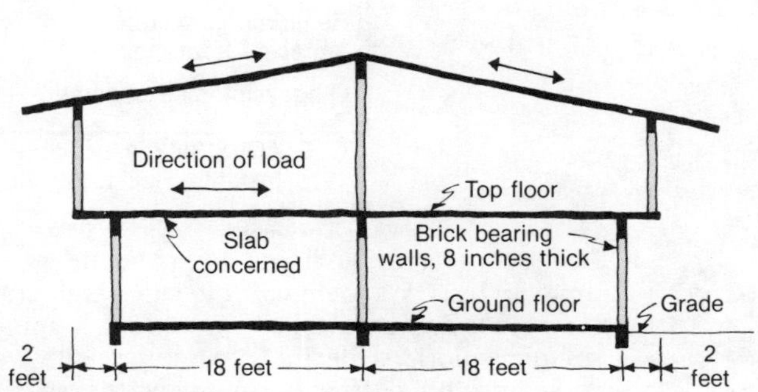

CROSS SECTION OF BUILDING

Since you do not give us any indication of the loads to be carried by the slab, we cannot give you a detailed recommendation. However, if this is a residence, it may be that live loads are relatively light and that the residual strength of the slab is adequate. We suggest that you consider a load test, during which you measure deflection of the slab to verify its performance. This approach may be the least expensive one in a relatively small structure and it should tell you much about the structural capacity that you have already in place.

Test for presence of fly ash

Q. *Is there a test for determining whether or not fly ash has been used in concrete? If so, can the test show how much was used?*

A. A petrographer can readily tell if fly ash is present in a sample of the hardened concrete. The exact amount is difficult to determine, but the petrographer can probably tell by visual methods if there was 1% to 2% or 10% to 15% fly ash by volume of cement. If samples of the fly ash used were available, a more precise estimate could be obtained using quantitative chemical analysis.

When to test for chloride ion?

Q. *The specification for our parking structure job says chloride ion content of the concrete should not be more than 0.15 percent by weight of the cement. We also had to submit certification of chloride ion content of the admixture, the coarse aggregate, and the fine aggregate. We have obtained the necessary certifications for the components of the mix and have a concrete specimen ready for testing. But WHEN should the test be done? We need to start placing concrete as soon as possible.*

A. "Building Code Requirements for Reinforced Concrete" (ACI 318-83) sets limits on water-soluble chloride ion content as a precaution against possible corrosion of reinforcing steel. The 0.15 percent limit called for by your specification applies to reinforced concrete exposed to chlorides during its service life. Limits for concrete under other conditions are shown in the table.

MAXIMUM WATER-SOLUBLE CHLORIDE ION CONTENT FOR CORROSION PROTECTION as required by "Building Code Requirements for Reinforced Concrete" (ACI 318-83)

Type of member	Maximum water-soluble chloride ion in concrete, percent by weight of cement
Prestressed concrete	0.06
Reinforced concrete exposed to chlorides in service	0.15
Reinforced concrete dry or protected from moisture	1.00
Other reinforced concrete	0.30

NOTE: ACI 318-83 is available from the American Concrete Institute, Box 19150, Detroit, Michigan 48219.

ACI 318 sets these limits for concrete at an age of 28 days but does not specify the time of test. We believe you can safely test before 28 days because research indicates that the soluble chloride content of a mix diminishes rapidly in the first 10 days. The soluble chlorides begin to combine with various ingredients of the concrete and become insoluble. If you can meet the specified limit at less than 28 days, the concrete will still be satisfactory (and maybe even better) at 28 days. *Concrete Construction* reported some background information on this in a special issue on the chloride question, October 1982, page 803.

Modulus of elasticity measures stiffness

Q. *A building owner wants to know the modulus of elasticity of our standard 4000-psi mix. What property does modulus of elasticity measure? Do I need a special test or can I calculate the value from strength test data?*

A. Modulus of elasticity is a measure of material stiffness or resistance to deformation. A concrete beam or column with a high modulus of elasticity won't deform as much under load as one with a low modulus of elasticity. For normal-weight concrete with a 4000-psi specified compressive strength, the modulus of elasticity is approximately 57,000 x $\sqrt{4000}$ = 3,600,000 psi. This is based on a relationship ($E_c = 57{,}000 \sqrt{f'_c}$) given in Section 8.5.1 of the American Concrete Institute Building Code (ACI 318-83).

To calculate modulus of elasticity based on actual concrete testing, you need to accurately measure deformation as well as load for a molded concrete cylinder or a drilled core. The procedure is described in ASTM C 469-87, "Standard Test Method for Elasticity and Poisson's Ratio of Concrete in Compression." Most testing laboratories are equipped to run this test.

Shorter cylinders have higher strengths

Q. *At our ready mix concrete plant we test cylinders for our own quality control purposes. I want to use neoprene pads on cylinder ends instead of capping compound. However, there's not enough crosshead movement in our testing machine to allow room for the cylinder plus pad. I thought of cutting off a half inch from the cardboard mold to solve the problem. However, I heard that cylinder length affects strength. How much would the strength be changed by reducing length by 1/2 inch?*

A. Not much. As the length-to-diameter ratio decreases, apparent strength increases slightly. To correct for this effect, the calculated strength is multiplied by a correction factor. ASTM C 42, Standard Method of Obtaining and Testing Drilled Cores and Sawed Beams of Concrete, gives the following correction factors for cores with strengths from 2000 to 6000 psi:

Ratio of cylinder length-to-diameter	Strength correction factor
1.75	0.98
1.50	0.96
1.25	0.93
1.00	0.87

The same factors can be used for cylinders. For your 11 1/2 inch-long cylinder, the length-to-diameter ratio is 1.92 and the correction factor is 0.99.

Field-cured cylinders

Q. *How closely do field-cured cylinders represent the strength of in-place concrete?*

A. Field-cured cylinder test results don't usually agree very closely with core test results or with results of other methods used to measure in-place strength. In one study, pairs of slabs made with concrete from the same batch were subjected to good and poor curing. Test cylinders were subjected to the same curing conditions. Cores were taken from the slabs and core test results were compared with cylinder test results. Cores from the well-cured slabs averaged about 10 percent less in strength than the well-cured field cylinders. And the strength of cores from poorly cured slabs averaged about 20 percent less than the poorly cured field cylinders. In other words, poor curing had a greater effect on the slab cores than on the cylinders. For this reason it could be dangerous to depend on field-cured cylinder strengths as a guide to safe form removal times.

Don't waste your pullout specimens

Q. *If you are using pullout tests to determine stripping time for formwork, how do you know when to pull out the specimens? If done too soon you might use up all the available pullouts and not have one available when you need to prove that it is safe to strip.*

A. A maturity meter that resembles a thermometer should help. You cast it into the concrete and read it periodically. This meter is really a device that gives what is called a time-temperature maturity reading. This value is a measure of how much, up to the time you read the meter, the temperature experience of that particular concrete has contributed to its strength. Then all you do is use that number, with a chart made for your particular mix, to read the approximate strength at the time. This tells you whether it is getting to be about time to make a pullout test.

Handles for test cylinders

Q. *Why don't compression test cylinders come with handles? They are awkward to carry, especially over rough terrain, and if you drop one you've lost your test data.*

A. Your question is timely because L. B. Shotwell, of Master Builders, has recently devised a homemade reusable handle that is described in "A Simple Device for Carrying 6x12-in. Concrete Test Cylinders," *Concrete International*, March 1983, pages 39 and 40. It consists of a 3/4x1x6-inch smooth piece of wood with a 1/4- or 3/8-inch hole drilled lengthwise through it and a 54-inch length of 1/4-inch-diameter polypropylene rope. Two pieces of 3/8-inch (outside diameter) flexible copper pipe, 1/2 inch long, are slipped over the rope and hammered flat to form permanent loops in the rope. These loops should be just slightly larger than the circumference of the 6x12-inch test cylinder mold.

Another test cylinder handle

In the May issue, page 427, a simple device was pictured for carrying 6 x 12-inch test cylinders in their molds. A reader has sent a description of another such device:

The device shown, made by one of our good technicians, is used to handle 6 x 12-inch concrete test cylinders in the field and in the laboratory. These are easily made,

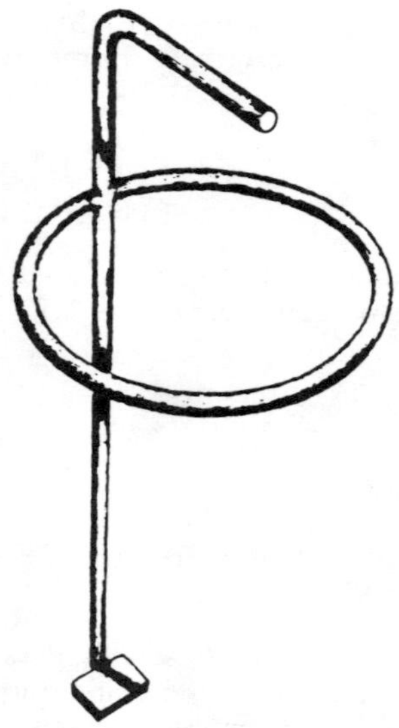

they can be put on the cylinder by one hand, and the technician can carry two cylinders at one time.

How to make saturated limewater

Q. *Test cylinders for an upcoming job are supposed to be cured on site. ASTM C 31 (section 7.3) says that after the cylinders are stripped they should be cured in saturated limewater. How much lime should I add to a 55-gallon drum of water to make a saturated limewater solution?*

A. Hydrated lime (calcium hydroxide) makes a saturated solution when less than 0.2 percent has dissolved. This is only 2 grams per liter or 0.9 pound per 55 gallons.

The definition of a saturated solution is one that contains as much solute (lime) as the solvent (water) will dissolve while in contact with excess solute. The simple answer to the problem is to thoroughly stir 2 to 5 pounds of hydrated lime into 55 gallons of water, allowing the excess to settle. The excess in the bottom will ensure that the solution remains saturated.

Accelerating tests of cylinders

Q. *For a long time we have considered taking the plunge and going in for accelerated strength testing. We know that cylinders tested in accordance with ASTM C 684 are broken at age 24 hours by the warm water method, at 28½ hours by the boiling water method, and at 49 hours by the autogenous curing method. No matter which method we adopt, our problem is how to be sure that we interpret the data properly so that we don't either accept strengths that are too low or require strengths that increase the cost unnecessarily. Can you suggest some guidelines?*

A. Perhaps you should study the way strength data were obtained and interpreted in the construction of the CN Tower in Toronto. Details on setting up a program are given in Peter Smith's 39-page article "Opening the Oyster of Concrete Quality," Stanton Walker Lecture, National Sand and Gravel Association and National Ready Mixed Concrete Association, 900 Spring Street, Silver Spring, Maryland 20910.

Since 1978 the U.S. Army Corps of Engineers has been using data from the warm water method to aid in control of concrete construction on civil works projects. Anyone planning to use any accelerated method should study the American Concrete Institute committee report, Accelerated Strength Testing, ACI 214.1R-81.

Correction factors for core strengths

Q. *When the testing laboratory broke some cores they calculated the strength in psi by dividing the load by the average cross-sectional area. Then they multiplied the result by a "strength correction factor" that gave a lower value. Where did this correction factor come from and how is it justified?*

A. The correction factor should have come from Section 5.7 of ASTM C 42, "Standard Method of Obtaining and Testing Drilled Cores and Sawed Beams of Concrete." The reason for applying a correction factor is that tall specimens tend to give lower test values that short specimens of the same cross-sectional area. In order to convert all core specimens to approximately the same strength basis as the standard 6x12-inch test cylinder, a correction factor is applied to the calculated strength. This factor, based on the ratio of length to diameter (l/d), has been given as 0.98 for $l/d = 1.75$, 0.96 for $l/d = 1.50$, 0.93 for $l/d = 1.25$ and 0.87 for $l/d = 1.00$.

Who should bear the cost of concrete replacement?

Q. *Our company supplied concrete for some flatwork from which 15 cubic yards had to be torn out and replaced because of low test strengths. We are not convinced that the low strength measurements were valid. How can we determine who should pay the cost of removing and replacing this concrete?*

A. In ASTM C 94, Specification for Ready-Mixed Concrete, Section 17 deals with such problems. It suggests that if the manufacturer and purchaser cannot agree, a panel be formed of three qualified engineers to adjudicate the controversy. One of these engineers is designated by the manufacturer, a second by the purchaser, and the third by the other two members of the panel. By prior agreement their decision is binding.

A useful six-step procedure for adjudicating such disputes is described in the document, "In-Place Concrete Strength Evaluation—A Recommended Practice," *NRMCA Publication Number 133-79*, National Ready Mixed Concrete Association, 900 Spring Street, Silver Spring, Maryland 20910.

Ages for strength testing

Q. *Why are strengths related to time periods of 7 days and 28 days?*

A. One could perhaps say that the reasons are historical—that these time periods were established in the early days of cement and concrete testing and have persisted ever since.

However, there seem to be good reasons that these time periods were chosen. One of the practical considerations in selecting testing ages is to avoid ages that fall on weekends. By selecting ages such as 7 and 28 days the test age will fall on the same day of the week as the day on which the test specimens were made—usually not a weekend.

On this basis, starting with multiples of 7 as being good ages, the question becomes *What multiples?* The curve in which compressive strength is plotted against age is generally of the form shown in the figure. As can be seen, enough strength has usually been attained by age 7 days to give some indication of whether the concrete is gaining strength at a satisfactory rate; so 7 days is a useful age at which to test. (Age 14 days is also sometimes used when an accurate record of the shape of the strength curve is needed.) Age 28 days gives a good clue to the strength that will be achieved by the end of the period of most rapid strength gain (see figure), so 28 days is a useful multiple of 7 days at which to test. Consequently, concrete mixes are usually proportioned in terms of the strength they can develop at 28 days.

Interestingly enough, now that high-strength concretes (6000 to 10,000 psi) have come into use, high-strength mixes are commonly proportioned to meet their design strengths at 56 or 91 days instead of 28 days, since the full strength is ordinarily not needed as early as 28 days.

One of the problems with testing at 28 days or later is that if the concrete does not meet the specified 28-, 56-, or 91-day strength it may have been a part of the structure for a long time before its deficiency becomes known. This problem has led to some reliance on accelerated methods of strength testing.

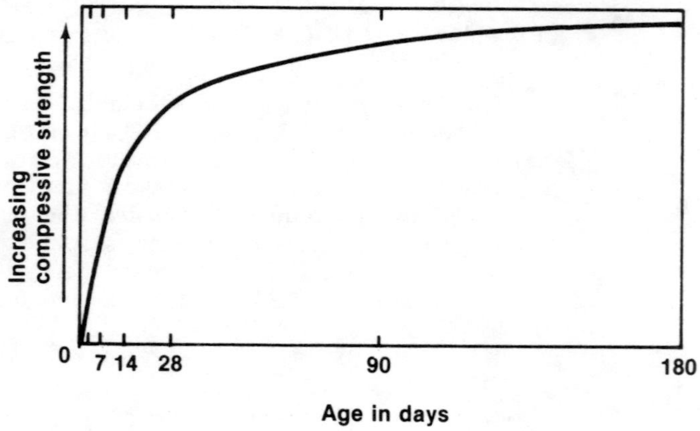

Identifying a problem through petrographic examination

Q. *We have what looks like a problem of durability with a concrete parking lot we put in last fall. Pitting of the surface showed up during the winter. At the time of construction the ready mix producer checked the air content and said it was all right. Now the architect wants a petrographic examination of the concrete to find the cause. What can we expect to learn from a petrographic examination and how sure can we be of the results?*

A. A petrographic examination of hardened concrete starts with visual examination of the concrete. Usually cores are removed from the concrete and sawed to expose a flat cross-section. A polished cross-section is then examined under a microscope. The visual analysis may be supplemented by X-ray or spectrographic examination as well.

In the visual examination, the laboratory checks the characteristics of the concrete most likely to be related to the problem. In your case that might include:

- Air entrainment—The total air will be checked through-out and examined at the surface to see if surface air was removed during finishing. As a contractor, the best procedure to follow during construction is to have your own air meter, preferably a pressure meter for normal concrete, and to make your own air tests at least twice during a day's pour. You can then prevent a problem caused by low air content for which your construction practices might be blamed.

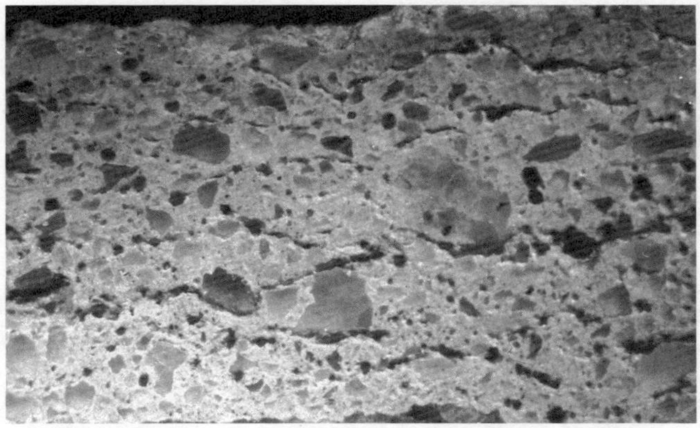

This cross-section view shows excessive mortar at the surface of improperly finished concrete. The veneer of mortar can be easily dislodged and is already breaking loose.

PHOTO BY: Erlin, Hime Associates, Northbrook, Illinois

- Density of the surface mortar—The more porous the mortar is, the wetter the concrete probably was at the time it was finished and the less durable the surface is likely to be. Other finishing characteristics might also be identified.
- Void space under the surface—The surface may have been sealed too soon, trapping bleed water beneath the surface.
- Aggregate soundness—Soft, porous, or alkali-reactive aggregate particles may be the source of pitting. Whether there would be a continuing problem of deterioration would be determined by identifying the character and the quantity of such material in the concrete mix.
- Frozen surface—This isn't likely in your case, but it can be identified if it is.
- Chemical attack—This is also unlikely, but is possible to determine.

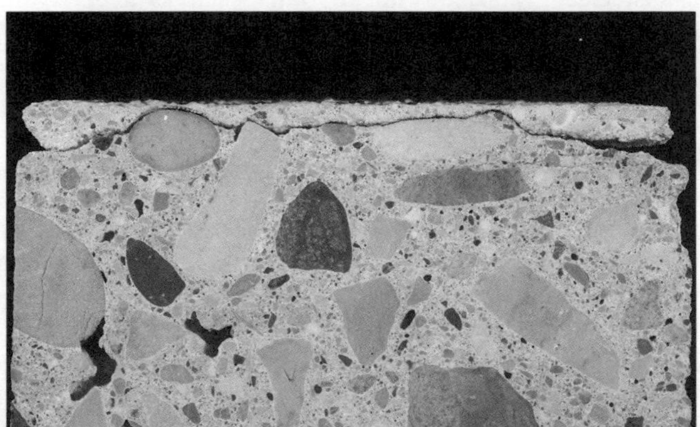

Cracks in the mortar caused by early freezing of green concrete are seen in this photomicrograph of a polished cross-section.

PHOTO BY: Law Engineering Testing Company, Atlanta, Georgia

Cement content may also be estimated, but probably not accurately enough to determine whether or not it contributed to the problem. The accuracy for this test can be improved if samples of the aggregates and the cement used in the concrete mix can be analyzed separately in addition to the examination of the hardened concrete. Comparing a concrete sample from an area which shows no distress with a concrete sample from an area that does can also be helpful.

You may find it useful to read American Society for Testing and Materials Standard Practice for *Petrographic Examination of Hardened Concrete,* ASTM C 856. Petrographic analysis is not as commonly understood as most other standard tests. For that reason ASTM C 856 states "It is assumed that the examination will be made by persons qualified by education and experience to operate the equipment used and to interpret the results obtained." You will want to be sure that the analysis is made by someone with that kind of experience and knowledge.

Petrographic examination provides information from which educated judgments may be made regarding the cause of a problem. Future performance of the concrete can sometimes be predicted. As it pinpoints the problem petrographic examination is also helpful in determining what action, if any, is necessary to make the concrete serviceable.

Measuring moisture in slabs

Q. *We have to provide a two-part protective membrane sealer on some slabs, both on grade and above grade. Sand will be broadcast into the sealer before it dries, providing a nonslip surface. The specification requires that the slabs be 28 days old before any sealer is applied. As usual we need to gain some time in the schedule, and if we can show a satisfactory relative humidity reading we may be able to start applying the sealer early. But how can we measure humidity? We used the plastic sheet test on a previous job and ran into trouble with the coating.*

A. A May 1986 article in *Concrete Construction* (page 472) described ways to determine when a slab is dry enough to apply coatings. We reported on several kinds of moisture meters that determine the relative humidity or moisture content of the concrete, including:

- Embedded meters
- Insert-meters cast in wells
- Surface meters

The surface meters are the simplest, requiring no damage to the concrete. One caution, however. Be sure that the measurement method you use is satisfactory to the supplier of your sealer material. The moisture content may be significantly less at or near the surface than at lower depths in the slab.

One of our readers, Clifton E. R. Lawson, also reported (July 1986, page 667) poor results with the plastic sheet test. He suggested a simple alternative: test cores or chips of concrete for moisture content by drying in an oven. Mr. Lawson also warned against cleaning slabs with anything such as a detergent solution that might leave salts in the concrete. He said this can sometimes result in water-filled blisters because the concrete can behave as an osmotic membrane.

Rate of evaporation from slab

Q. *For an electronics company operation we recently designed a slab on grade built on a moisture barrier over a gravel base to receive a special antistatic vinyl flooring. Warranty requirements for the vinyl were that the rate of emission of water vapor from the floor be not more than 3 pounds per 1000 square feet per 24 hours. When the slab was approximately 6 months old (the building having been enclosed for 3 months and heated for about 2 weeks) tests gave a value of 6 pounds per 1000 square feet per 24 hours.*

The test consisted of placing a petri dish containing anhydrous calcium chloride on the slab and covering it with a 0.5-square-foot plastic cover. The cover was sealed to the surface to prevent moisture from being taken up from the air of the room. The dish of calcium chloride was removed after 24 hours and weighed to determine how much moisture had come out of this area of the floor.

The owner, contractor, and design team believe that the moisture criteria may be excessively stringent. What kind of emission of moisture vapor can be expected from a curing concrete slab at 6 months? Do the test-measured moisture values correspond in any way to these values?

A. H. W. Brewer studied the rates of moisture emission from slabs on grade under a considerable variety of conditions and reported his results in "Moisture Migration—

Concrete Slab-on-Ground Construction," Portland Cement Association, *Development Department Bulletin D89*, published in 1965. (Available from Order Processing Department, Portland Cement Association, 5420 Old Orchard Road, Skokie, Illinois 60077.) All of his specimens were stored in a room at a constant temperature of 73° F and a constant relative humidity of 30 percent. Some specimens were cast in a container on a 4-mil polyethylene vapor barrier. During this particular test, water was kept in contact with the lower side of this vapor barrier. At an age of 180 days, according to Brewer's Figure 16, the rate of moisture outflow was about 16 grains per square foot per hour, which is equivalent to 2.33 pounds per 1000 square feet per 24 hours. It should be noted that these slabs were 4 inches thick, so the rate of outflow from them would be smaller than from thicker slabs. You may wish to study the rates of outflow under various conditions and at various ages as given in Brewer's report.

It should also be noted that your method of test may have tended to make the results high. Anhydrous calcium chloride is a strong desiccant. It attempts to maintain a relative humidity that is close to zero in the space immediately surrounding the crystals. This would mean that concrete that had reached almost a steady rate of moisture loss at the ambient relative humidity of your building would suddenly be exposed to a lower relative humidity. This new difference in partial vapor pressure between the concrete and the air under the plastic cover would immediately speed up the rate of moisture outflow. This would produce a higher apparent rate than the floor actually experiences at ambient relative humidity.

Showing whether moisture comes from above or below

Q. *In South Florida we have a recurrent problem with moisture on top of the slabs in houses and warehouses. We have devised a simple test that consists of using paper towels to dry up the water on top of the slab, then chipping off some concrete to expose material below the finished surface. We always find the concrete to be dry and that the dust from the chipping is dry when rubbed between our fingers.*

We have also done another kind of test. We drill cores from the slab and sample the fill under the slab as well as the fill outside the house. Our results always indicate the moisture content of the fill under the slab to be lower than the fill outside the house by several percentage points.

We are sure that this is water that has condensed on the slabs from the air when the humidity is high (79 to 85 percent in spring and summer). We are convinced that it is not water that has percolated through the slab, but we have trouble convincing our customers. Can you publish something regarding this problem and its solution?

A. We agree that your tests do show that the moisture is coming from the air, not from the subgrade.

There is another possible test that ought to be convincing. If you were to lay a piece of sheet metal, about two or three feet square, on the surface of the concrete and tape down the edges securely and leave it overnight, moisture should condense on the top surface of the metal if the air is cool and humid. When you then remove the tape and look at the underside of the sheet metal there should be no moisture on the bottom. This should convince the observer that the moisture is coming from above and not from below, since moisture would not go through sheet metal.

Under some circumstances you could get a small amount of condensation on the bottom of the sheet metal also, provided the slab had not yet dried out thoroughly and moisture were still coming from below. If so, you would have to agree that there is still moisture in the slab. In many cases, however, the top could be very moist while the bottom contains little or no moisture.

Tests on fresh concrete

Detecting the presence of fly ash in fresh concrete

Q. *Our specifications for some projects prohibit the use of fly ash. We have reason to believe, however, that fly ash has sometimes been used in violation of the specs. Is there any fly ash test that can be made quickly enough on fresh concrete to permit rejection of a delivery?*

A. If there is unauthorized fly ash in the mix it doesn't always mean it was put there purposely. Sometimes fly ash gets into a mix accidentally through a leaky wall between adjacent cement and fly ash bins or silos. At other times a small amount of fly ash may have got hung up in a chute that was used for batching fly ash for the previous mix and then this may break loose in a batch where it isn't wanted.

Whether fly ash has been purposefully added or not, there is sometimes a need to test for its presence and the Iowa Department of Transportation has worked out a successful test that can be used at the jobsite. It requires a microscope of about 60X to l00X, glass slides, U.S. standard sieves of Number 200 and Number 325 sizes, a plastic squeeze bottle containing water and one containing alcohol.

The procedure begins by nesting the sieves with the Number 325 sieve on the bottom. The operator places about one tablespoonful of the mortar portion of the concrete on the Number 200 sieve and washes it thoroughly, using the squeeze bottle containing water. This causes most of the sand to remain on that sieve while a portion of the cement and fly ash collect on the Number 325 sieve below it. The latter sieve is drained while being tapped lightly. Then the residue on this sieve is washed, using the squeeze bottle containing alcohol, until the liquid coming through is clear. After the residue has been allowed about 15 minutes to dry, a small portion is tapped from the sieve onto a microscope slide and examined, preferably with a dark background. If any glassy spheres can be seen it indicates that there is fly ash in the sample.

Further details and complete procedures are given in a publication of the Iowa Department of Transportation. One method permits taking a sample at the site, placing it in a jar of water and shaking it, and then transferring it to the laboratory for testing at some later time.

Copies of the procedure, "Method of Test for Determining if Fly Ash Is Present in Plastic Portland Cement Concrete or Portland Cement," *Materials Instructional Memorandum 371,* are available from the Office of Materials, Iowa Department of Transportation, 800 Lincoln Way, Ames, Iowa 50010.

Test for slump at the truck or at the end of the pumpline?

Q. *If concrete is to be pumped and the specification calls for a 4-inch maximum slump, should the slump test be taken at the point of discharge from the truck or at the point of discharge from the pumpline?*

A. Concrete sampling procedures are described in the ASTM standard method of Sampling Freshly Mixed Concrete (ASTM C 172). A note in this method says that sampling should normally be performed as the concrete is delivered from the mixer to the conveying vehicle used to transport the concrete to the forms. The note further states, however, that specifications may require other points of sampling, such as at the discharge of a concrete pump.

The point of sampling may depend on the purpose of the test. If the purpose is to verify the quality of the concrete as delivered, sampling from the mixer discharge is appropriate. Sampling at both the truck discharge and the point of final placement may be employed to determine if any changes in slump or other mix characteristics occur. To avoid disputes about where the sample is to be taken, put this subject on the agenda for a pre-job conference. The engineer, concrete producer, pumper, and testing laboratory representative should agree on sampling location *before* the job starts.

Make it routine: check the air

Q. *We built a 2-acre parking lot for a local shopping center. It was salted regularly during the first winter and by spring there was widespread scaling. The owner had some tests made that showed that the air content was very low in the scaled areas. We are surprised because we ordered air-entrained concrete. We've made a practice of having the air content checked twice each year by an admixture manufacturer and we've never had this kind of trouble before. Does this mean we can't count on the tests the manufacturer runs for us? Is there something else we should be doing to avoid such trouble?*

A. You can't count on one air content test to guarantee the air content in all the mixes you buy over several months. You can't even be sure that the air measured for one truckload will represent the air content in the next delivery, though the chance would be better.

When you're constructing outdoor flatwork that will be exposed to severe weathering, and particularly if deicers will be used on it, it's worth testing the air content at least daily. Some would say that make sure the air content is adequate is even more important to the life of such flatwork than making sure the concrete meets strength requirements.

The cost of air test meters is not high compared with the value of the information they provide. It doesn't take long to learn to use them, though it is true you must be sure to use them properly. The test only takes about 5 minutes. We think every flatwork contractor in severe and moderate (not mild) climates should use them regularly. An article that provides further introduction to air meters is "Every Flatwork Contractor Should Own an Air Meter," published in our September 1986 issue, page 801.

Testing for air content

Q. *Is there a practical way to test air content on the job?*

A. The standard test for air content used by inspectors and testing laboratories is relatively easy to learn to do properly, and, once learned, it only takes a few minutes to make the test. Complete instructions are given in ASTM C 231, "Standard Test Method for Air Content of Freshly Mixed Concrete by the Pressure Method," which appears in *Part 14, 1981 Annual Book of ASTM Standards* and also in Volume II of *ASTM Standards in Building Codes.* The Corps of Engineers, U.S. Army, publishes the same standard under its own designation "CRD-C 41-81," which appears in its *Handbook for Concrete and Cement, Volume I.*

An air test meter meeting the ASTM C 231 requirements can be purchased for somewhere in the range of \$500 to \$800 at current prices. If you are interested in acquiring and using one you should practice making the test on concretes that are being tested at the same time by a proficient and reliable operator. In this way you can learn to make the test accurately within the reasonable amount of operator error. Once you have learned to use the equipment properly you can make tests easily on the job and increase your confidence in knowing whether you are getting what you order.

Evaluating durability of a structure

Q. *As we are more and more concerned with durability, how do we measure it on a given structure after a year or after 5 years?*

A. If you are thinking about both surface condition and structural integrity you should consult the ACI 201 Standard, "Guide for Making a Condition Survey of Concrete in Service," which appears in Part I of the *ACI Manual of Concrete Practice.*

Illustrations from that report are reprinted in a series of articles in *Concrete Construction* magazine, "Surveying Concrete in Service," July 1969 pages 257 to 259, August 1969 pages 300 to 302, and September 1969 pages 347 and 349 to 352.

Nondestructive testing and petrographic examination can be run, if necessary, by some of the agencies listed in the professional services section on the next to the last page of current issues of *Concrete Construction.*

Index

Concrete

Part II: Masonry

Materials

Masonry

I

Ashlar masonry defined

Q. *I have bought an old house and am putting in a fireplace. A contractor recommended an ashlar masonry fireplace to match the rustic look of the house. What is ashlar masonry?*

A. Masonry terms can sometimes be confusing and meanings may differ in different parts of the country. One of the best dictionaries for masonry terms is *The Masonry Glossary,* published by the International Masonry Institute. Their definition for ashlar masonry includes two subgroups described below.

ashlar masonry—Masonry composed of variable size rectangular units having sawed, dressed, or squared bed surfaces, properly bonded, and laid in mortar.

coursed ashlar—Ashlar masonry laid in courses of stone of equal height for each course, although different courses may be of varying height.

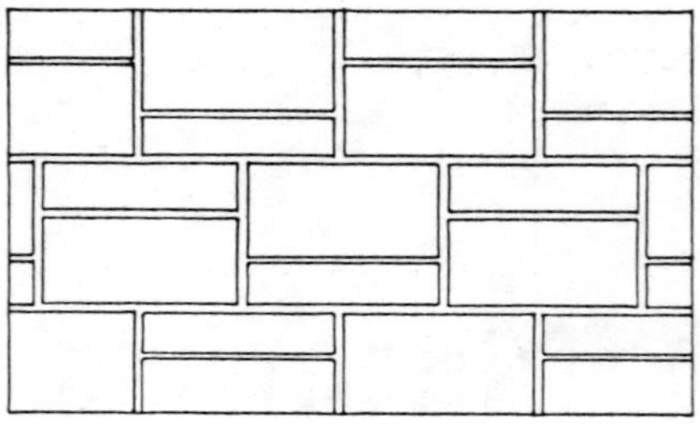

random ashlar—Stone masonry patterns of rectangular stones set without continuous joints and laid up without drawn patterns. If composed of material cut to modular heights, discontinuous but aligned horizontal joints are discernible.

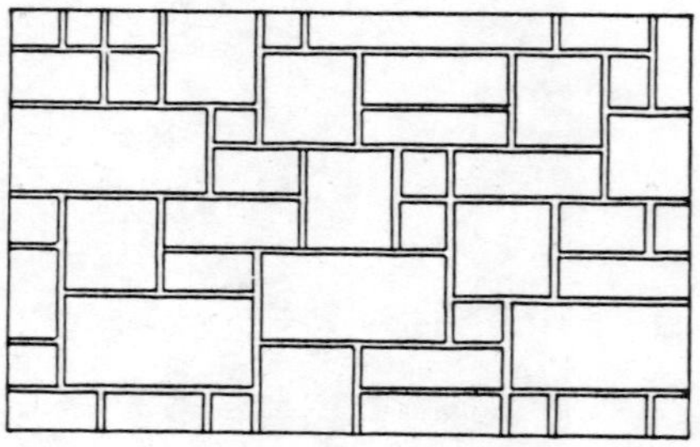

The Masonry Glossary can be purchased from the International Masonry Institute, 823 Fifteenth Street, NW, Washington DC 20005.

Chipped brick—who's responsible?

Q. *I just received a shipment of brick from my supplier. The faces of the brick are chipped, and I don't think the architect will approve the finished wall if I use the brick. If they meet specifications, am I stuck with the brick?*

A. If the brick meets the architect's specifications, the architect is stuck with the brick—even if he is not pleased with the exposed surface. ASTM C 216, Standard Specification for Facing Brick, allows different amounts of chipping for the different types of brick (see table). Compare the amount of chipping your brick has with the amount allowed by C 216 for the type of brick specified by the architect.

If the supplier's brick are not chipped more than what is allowed by C 216, he does not have to supply new brick. If the architect wanted less chippage, he should have specified a brick with a higher chippage tolerance or he should have specified Type FBA brick and the exact chippage tolerance he would permit. Explain this to the architect so he gets the exact appearance he wants in the future. In fact, before you order the brick, it may be a good idea to make the architect aware of the amount of chippage allowed on the type of brick he specified. This could save an argument in the field and a displeasing finished wall.

If the supplier's brick are chipped more than what is allowed by C 216, he should supply new brick or sort out the chipped brick that exceeds the allowed amount.

If you mistakenly ordered the wrong type of brick, you'll have to determine the least expensive way of getting the correct type of brick. You might be able to sort out the badly chipped brick. Or through talks, you might be able to come to an agreement with the supplier. Suppliers usually don't want to lose a contractor's future business.

Whatever the result, you're wise to be questioning the compliance of the brick before you place it. According to ASTM C 216, once the brick is in place the supplier is no longer responsible for its compliance. The contractor is.

BRICK CHIPPAGE ALLOWED BY ASTM C 216

Type of Brick	% of Exposed Brick that Can Be Chipped the Amount Listed	Inches of Chippage from	
		Edge	Corner
FBX	95% to 100% 5% or less	0 to 1/8 1/8 to 1/4	0 to 1/4 1/4 to 3/8
FBS smooth[1]	90% to 100% 10% or less	0 to 1/4 1/4 to 5/16	0 to 3/8 3/8 to 1/2
FBS rough[2]	85% to 100% 15% or less	0 to 5/16 5/16 to 7/16	0 to 1/2 1/2 to 3/4
FBA	As specified by the purchaser.		

[1] Smooth texture is the unbroken natural die finish.

[2] Rough texture is the finish produced when the face is sanded, combed, scratched, or scarified or the die skin on the face is entirely broken by mechanical means, such as wire-cutting or wire-brushing.

Source: ASTM C 216, ASTM, 1916 Race Street, Philadelphia, Pennsylvania 19103.

Is an IRA of 3 too low?

Q. *In several recent projects, the owner selected a brick with an IRA less than 3. Can sufficient bond be achieved with this dense of a brick?*

A. IRA (initial rate of absorption) is a measure of how much water a brick absorbs on the surface in 1 minute. As a general rule, the highest bond occurs with the brick having IRA values between 5 and 30 grams per minute per 30 square inches of brick surface. Most brick with IRA values less than 5 can achieve sufficient bond. When combined with some mortars, however, some low IRA brick have developed lower than normal bond because the brick tend to float on the mortar. Before using a brick with a low IRA, consider performing bond strength tests on prisms made with the brick and mortar proposed for the project. Other factors, such as the roughness of the brick surface and the mortar type, also influence brick-mortar bond.—*Norbert V. Krogstad*

C 270 no longer specifies Type K mortar

Q. *What is Type K mortar? I have a job where the specifications require Type K mortar, but I can't find information about the mortar in any of the standards.*

A. Welcome to the changing world of specifications. Type K mortar has been eliminated from ASTM C 270 Standard Specification for Mortar for Unit Masonry. C 270 now contains proportion and property specifications for only Types M, S, N, and O mortars.

Politely point out to the designer that Type K mortar no longer has a proportion or property specification. If the mortar is intended for interior nonload-bearing masonry, which is what C 270 used to recommend Type K mortar for, suggest using Type O mortar instead. If the mortar is to be used for other uses, refer the designer to Table XI, Guide for the Selection of Masonry Mortar, in ASTM C 270. Also refer him or her to the article, "How to Specify Mortar," by Colin C. Munro, on page 22 of the April 1988 issue of *The Magazine of Masonry Construction.*

Does retempering mortar reduce strength?

Q. *I'm confused. I am a small-sized subcontractor that contracts for work in both concrete and masonry. Most inspectors don't let us add more water to concrete. For masonry mortar, however, some inspectors let us retemper with water and some don't. I know that adding extra water to concrete lowers its strength. Doesn't extra water also lower the strength of mortar?*

A. You are correct about concrete. Adding extra water onsite increases the water-cement ratio and decreases strength. A good rule of thumb to remember is the addition of 1 gallon of water for every bag of cement decreases the compressive strength by 1000 psi.

Masonry mortar and concrete are similar materials but are used quite differently. However, because of their similarities, sometimes specifications for concrete are incorrectly applied to masonry mortars.

Mortars differ from concrete in method of placement, workability, and structural use. Concrete is used as a single structural element; mortar is used to bind other structural units together. Consequently, mortar bond strength is more important than its compressive strength. It's the bond of the mortar to the masonry units that controls the failure of most masonry.

If water in the mortar mix evaporates or is absorbed by the masonry units, the mortar-unit bond is reduced. Highest bond strengths are obtained with moist mixes having good workability. Retempering, the addition of mixing water to compensate for water loss, is an acceptable practice for masonry construction. There are, however, some requirements for this practice. ASTM standards require that all mortar be used within 2½ hours but permits retempering as frequently as needed within this time period. Tests have shown that the decrease in compressive strength due to additional water is minimal if retempering occurs within 2½ hours. Mortar that is not used within 2½ hours or that has begun to set should be discarded. Also, the retempering should not be accomplished by simply splashing water on the mortar surface. This practice tends to wash the fines from the mortar. Instead, form a basin in the mortar, add the water, and then thoroughly mix.

Some inspectors probably rely on their concrete experience to assess masonry construction. Refusing to allow retempering results in mortars that have maximum compressive strength, but less than optimum bond strength.

Colored mortar concern

Q. *Water penetration problems are occurring in the dark-colored mortar joints of a masonry building we designed. Is this because the joints are colored? Does water penetrate colored mortars more than it does uncolored mortar? Is this because the pigments add more non-cementitious material to the mix, reducing mortar strength?*

A. From the information you give, we can't tell you what's causing the water penetration, but it's probably not because the mortar is colored. Color pigments do not cause problems unless too much of them are added. The maximum pigment dosage recommended by most pigment manufacturers is 10% by weight of cement. Pigments that contain carbon black should be added in dosages no greater than 2% by weight of cement.

All the pigment manufacturers we contacted said that color pigments tend to fill the voids in the mortar, making it less permeable. Compressive strength is greater, and tensile strength is unaffected.

Pigments do add more non-cementitious material to the mix, but the relative amount is small. Typically, 5 pounds of pigment are added to a mortar batch containing 200 to 250 pounds of sand—about a 2% increase in non-cementitious material. Given the relatively inaccurate way mortars are made on the jobsite, this increase in noncementitious material is meaningless.

Good mortar color can be obtained with recommended dosages if the pigment is added in the proper manner. Generally, all or a part of the aggregate should be added to the mortar mixer first. Then the pigment and some water is added. The mixer is run for a few minutes, the cement and the rest of the aggregate and water are added, and then all the materials are mixed for at least another 5 minutes. These mixing rules don't apply if precolored masonry cement is used. But in either case, each batch of mortar should be batched in exactly the same way and mixed for the same amount of time. Batching inconsistencies produce variations in mortar color. Also, colored mortar should not be retempered. The extra water can change the color of the mortar.

Problems can arise if excessive amounts of color are added. The excess amount of fine pigment particles demand more water, which increases the water-cement ratio of the mix and decreases strength. Lower strength mortar deteriorates quicker, and thus is more susceptible to water penetration. But unless the contractor added more than 15% pigment by weight of cement (against manufacturer recommendations), color is not the cause of your problems.

Is it the test procedure or the test result?

Q. *Our job has just been shut down because the mortar doesn't meet the specified strength. The architect specified Type M mortar (2500 psi) under ASTM C 270, Specification for Mortar for Unit Masonry. The testing laboratory molded mortar cubes on the job from samples taken from the mixer, then took the cubes to the laboratory for curing and testing. The engineer at the testing laboratory told us the tests were done according to ASTM C 109, Test Method for Compressive Strength of Hydraulic Cement Mortars, as directed by C 270.*

The architect has called a conference to resolve the problem. We think he'll require us to add more cement to the mortar. Because of the hot weather, we're already having trouble with short boardlife and are afraid more cement will make the problem worse. What other alternatives can we suggest?

A. You're caught in the middle of a misunderstanding about ASTM C 270. Mortar that meets the requirements of C 270 must be made in the laboratory using the same materials and proportions that will be used on the job. The entire C 109 testing procedure—proportioning, mixing, molding, curing, and testing—must be done in the laboratory. Mortar cubes made in the field and tested in the laboratory do not have to meet C 270 requirements.

You can test job-mixed mortar using ASTM C 780, Preconstruction and Construction Evaluation of Mortars for Plain and Reinforced Unit Masonry. But results of ASTM C 780 tests also don't have to meet strengths specified in C 270.

At the moment, the project architect has no qualified test data to help him determine whether or not the mortar meets specifications. Here are two suggestions you could make at the meeting:

- Require that laboratory tests be made on laboratory-prepared mortar using materials and proportions to be used on the job. If the results meet the requirements of ASTM C 270, the mortar is acceptable.
- Require that C 780 tests be made on job-mixed mortar and their results correlated with results from laboratory tests on laboratory-prepared mortar. Correlated C 780 results would then be a guide for determining if mortar is acceptable. Be aware that the correlation may change with changing job conditions.

Making prehydrated mortar

Q. *The specifications for the project I am working on require prehydrated mortar. Is a mortar that uses prehydrated lime considered a prehydrated mortar?*

A. No. A prehydrated mortar is a mortar that is mixed 1 to $1^1/_2$ hours before it is used. Prehydrated mortar is used for tuckpointing to reduce the amount of shrinkage. Mortar that shrinks less creates tighter joints. Letting the mortar sit awhile can reduce shrinkage because a lot of water is used in the chemical reactions that occur in the first 1 to $1^1/_2$ hours. Using prehydrated lime doesn't make mortar a prehydrated mortar. Regardless of their use, all portland cement-lime mortars should be made with prehydrated lime. Technical Note 7F, published by the Brick Institute of America (BIA), describes how to make a prehydrated mortar: Thoroughly mix all dry ingredients, then add only enough clean water to produce a damp, workable consistency that will keep its shape when formed into a ball. Let the mortar stand in this dampened condition for 1 to 1½ hours.—*Norbert V. Krogstad*

Insulating 8-inch block

Q. *Can you recommend a method of insulating regular 8-inch concrete masonry units that performs as well as an insulated 6-inch wood-frame house and is still cost competitive? Would rigid insulation on the exterior and veneer or lathe and stucco do the trick? I need some statistics to show comparisons. My area is in the White Mountains of Arizona: 7,000 feet elevation, -20° F to +88° F. This may help us get into the housing market in this area.*

A. Following the detail in the chart shown here, you can build a concrete masonry wall with an R value equal to the R value of an insulated 6-inch wood-frame wall. Check with your local material suppliers for exact R values and prices of materials. Using molded expanded polystyrene or extruded expanded polystyrene instead of expanded polyurethane would provide R values of 7.00 or 10.00 instead of 12.50.

INSULATION NEEDED TO GIVE 8-INCH CMU WALL SAME R VALUE AS INSULATED 6-INCH WOOD-FRAME WALL

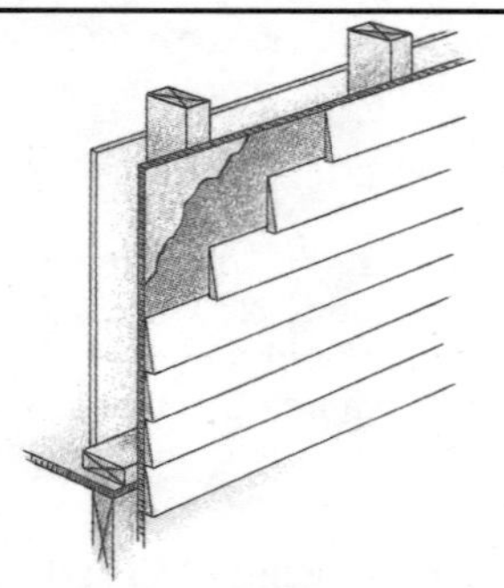

6-inch wood-frame wall	R value
1. Outside surface film	0.17
2. Wood siding	0.81
3. Sheathing felt	0.06
4. Plywood sheathing, ½ inch	0.62
5. Batt insulation	19.00
6. Gypsum plasterboard, ½ inch	0.45
7. Inside surface film	0.68
Total	21.79

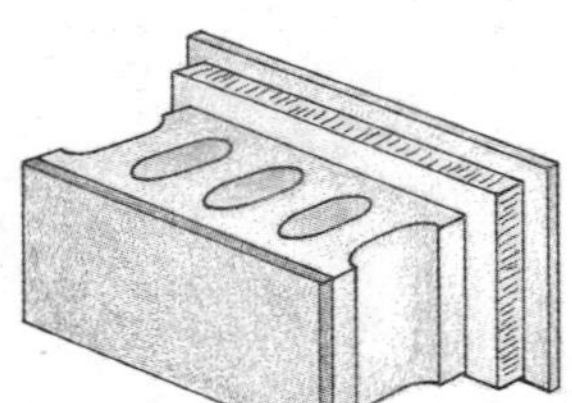

8-inch concrete masonry wall	R value
1. Outside surface film	0.17
2. Stucco, ¾ inch	0.15
3. Lightweight concrete masonry	2.12
4. Cores filled with insulation	4.85
5. Expanded polyurethane rigid insulation, 2-inch	12.50
6. Gypsum plasterboard, ½ inch	0.45
7. Inside surface film	0.68
Total	20.92

Sources

Means, R. S., *Building Systems Cost Guide,* 100 Construction Plaza, Kingston, MA 02364.

Olin, Schmidt, and Lewis, *Construction—Principles, Materials and Methods,* Institute of Financial Education, 111 E. Wacker Dr., Chicago, IL 60601.

"Thermal Insulation of Concrete Masonry Walls," NCMA-TEK 38-A, National Concrete Masonry Association, 2302 Horsepen Rd., Herndon, VA 22070.

Estimating loose-fill insulation

Q. *I am estimating a job that calls for loose-fill insulation to be placed in hollow concrete masonry walls. How do I calculate the amount of insulation needed to fill the cores?*

A. This table lists how much insulation is needed for various types of masonry walls. About 28 cubic feet of loose-fill insulation are needed to fill 100 square feet of an 8-inch hollow lightweight concrete block wall. The insulation triples the R value of the wall.

VOLUME OF INSULATION NEEDED TO FILL CORES OR CAVITIES OF MASONRY WALLS

Type of Wall		Cubic Feet of Insulation Per 100 Square Feet of Wall
Cavity Wall	**Width of Cavity (inches)**	
	2	16.7
	2½	20.0
	4	33.4
Single-wythe Hollow Concrete Masonry Wall	**Wall Thickness (inches)**	
	6	19.0
	8	27.6
	10	35.0
	12	50.0

Source: "Estimating Concrete Masonry," NCMA-TEK 14, National Concrete Masonry Association, 2302 Horse Pen Road, Herndon, Virginia 22070.

C/B ratio for brick

Q. *I have been in the masonry business for some time but was recently stumped when asked for the C/B ratio for our brick. What is a C/B ratio?*

A. The C/B ratio is short for cold water/boiling water absorption ratio, which is another name for the saturation coefficient. It's an indicator of brick's resistance to freezing and thawing.

To calculate the saturation coefficient, cold water absorption and boiling water absorption are measured. Samples of the brick are oven dried and weighed. Then they're submerged in cold water for 24 hours, surface dried with a damp cloth, and weighed again. The weight of water absorbed is divided by the oven dry weight and expressed as a percentage. The same samples are next submerged in water, heated to boiling, and boiled continuously for 5 hours. After surface drying with a damp cloth they're weighed again. The weight of water absorbed during boiling also is divided by the oven dry weight and expressed as a percentage. The saturation coefficient is the ratio of cold water absorption to boiling water absorption. The test method is more completely described in ASTM C 67, Standard Methods of Sampling and Testing Brick and Structural Clay Tile.

The saturation coefficient is almost always less than 1.0, and the lower the value the less absorptive the brick. ASTM C 62, Standard Specification for Building Brick, sets maximum permissible values for brick that will be exposed to freezing and thawing.

However, these requirements may be waived if the brick passes a freezing and thawing test, also described in ASTM C 67.

C/B ratio: an indicator of freeze-thaw resistance

Q. *Joseph F. Edwards from General Shale Products Corp. had additional information on "C/B ratio for brick."*

A. Your answer to the question, "What is a C/B ratio?," in the September 1988 issue contains an incorrect statement that is a common misconception. You state in paragraph three that the lower the value of the C/B ratio, the less absorptive the brick. A C/B ratio of 0.50 can exist in a very absortive brick (40% Cold Water Absorption/80% Boiling Absorption) or in a practically nonabsorptive brick (0.1% Cold Water Absorption/0.2% Boiling Absorption). C/B ratio has no bearing whatsoever on the total absorption of a brick. It represents roughly the ratio between the pores in a brick that can be filled by normal temperature and pressure water in a 24-hour period and the total pores in the brick.

According to the ASTM C 216* specification for facing brick, Grade SW (Sever Weather) brick cannot have a C/B ratio greater than 0.78. This maximum value was derived from thousands of weathering tests done by the National Bureau of Standards in the 1930s. It assumes that 78% of all the pores that are available in a brick can be reached by ordinary rain water and capillary action. The remaining 22% provide a place of escape for water when it freezes and expands. Acting as an expansion joint, these remaining pores prevent a saturated brick from shattering when absorbed water turns to ice. Therefore, the C/B ratio represents a brick's potential freeze-thaw durability, not its total absorptiveness. According to the above theory, the lower the C/B ratio of the brick, the more resistant to freeze-thaw it should be.

*Standard Specification for Facing Brick (Solid Masonry Units Made from Clay or Shale), ASTM C 216, ASTM, 1916 Race St., Philadelphia, PA 19103.

Lightweight or heavyweight block?

Q. *Our project calls for lightweight concrete block. Aren't heavyweight concrete block stronger and better to use?*

A. Heavyweight block may be slightly stronger than lightweight block, but ASTM C 90, Standard Specification for Hollow Load-Bearing Concrete Masonry Units, requires both types of block to have the same minimum strength. And both types of block can be used in any type of construction.

Lightweight units are more fire resistant, more sound resistant, and provide more insulation. Because they're lighter, they also are easier to lay, which means masons can lay more of them in a day—as much as 10% more.

However, because lightweight block are more porous than heavyweight block, they may be more susceptible to water penetration and freeze-thaw damage. This is perhaps why in the Northeast and Great Lakes states, heavyweight block are used more than lightweight block. Nationally, lightweight block are used as much as heavyweight block.

Copper flashing that has turned black

Q. *Our firm is involved in a project where the through-wall masonry flashing is copper. When a 2-day-old portion of the wall was taken down to correct a problem with the coursing, I noticed that the copper flashing had turned black wherever it was in contact with mortar. Is this a problem?*

A. No. The black color is a copper oxide that forms in the high pH environment of mortar. After this black oxide forms, there is essentially no further corrosion. The copper flashing is then stabilized and protected by the portland cement in the mortar.—*Courtesy Norbert V. Krogstad*

Construction and Repair

2

Masonry

Weather-resistant joints

Q. *For aesthetic reasons, we'd like the brick building we're designing to have raked mortar joints. Even though we've seen buildings with raked joints in good condition, we recently heard that raked joints don't weather well. Is this true? What styles of mortar joints weather best?*

A. The style of mortar joint does greatly influence the watertightness of a wall. Concave and vee joints are the most watertight and weather-resistant. Beaded and weathered joints also perform satisfactorily. Raked, extruded, struck, and flush joints, however, are not weather-resistant and should not be used on exterior surfaces.

Materials and workmanship also influence the watertightness of mortar joints. Mortar should meet the requirements of ASTM C 270, Specification for Mortar for Unit Masonry. Brick generally should have an IRA in the range of 5 to 30 grams per 30 square inches per minute. Brick with an IRA greater than 30 grams per 30 square inches per minute should be wetted.

Masons should set the units within 1 minute of spreading mortar, and they should not spread mortar beds more than 4 feet in front of the masonry. If they spread more mortar or take longer to set units, the wind and sun can dry the mortar before the units are placed. This creates a weak mortar-brick bond that allows water to penetrate. Masonry units also should never be moved after they are laid; any slight movement can break the bond between the unit and mortar.

The mortar joints should be tooled when the mortar is thumbprint hard, not sooner or later. Tooling pushes the mortar back into contact with the masonry units, closing slight hairline cracks that form as the mortar hardens and shrinks. It also compacts the surface of the mortar. If the tooling is done too soon, the mortar only shrinks away from the masonry units again. If it's done too late, the mortar is too hard and can't be pressed tight against the masonry units.

Because exterior joints must be watertight, all head joints should be filled, even if they are not exposed. And they should be buttered, not filled by slushing.

Some factors that affect the watertightness of mortar joints are described in more detail in "The Most Important Property of Masonry," by Clayford T. Grimm, published in the September 1988 issue of *The Magazine of Masonry Construction*, page 254.

Why wide joints are wrong

Q. *As this photo shows, the exterior wall of our building has several cracks in the wide mortar joints. The contractor evidently made all joints the same width as the extra-wide joint containing the 3/8-inch-thick shelf angle (the dark slot shown in photo). Do wide mortar joints inevitably suffer cracks?*

A. Masonry consultant Colin C. Munro of Batavia, Illinois, says cracks are much more likely to occur in very wide mortar joints. Joints typically should be $^3/_8$ to $^1/_2$ inch wide, depending on the size of the brick. To lay much wider joints, the mason must use a stiffer mortar, probably one that contains less water and less lime. The result: not only is there more mortar (and more mass), but there's also a greater proportion of cement in the mortar. This high-cement mortar undergoes more drying shrinkage and thus more shrinkage cracks occur. The problem is worst when brick with a high initial rate of absorption (IRA) also is used. The brick will absorb water, making the mortar even drier.

Don't forget to install horizontal soft joints

Q. *We completed a 10-story building about 1 year ago. The building frame is concrete and the exterior is brick veneer. The exterior face brick recently cracked and disintegrated at the first floor window heads (see photo). The same thing is beginning to happen at the second and third floors. The problem is occurring only at corners and piers that are continuous the full height of the building. It looks like the brick is being crushed under its own weight. What is causing this and how do we fix it?*

PHOTO CREDIT: Turner Smith, Jr.

A. The photograph does not show any horizontal soft joints. That is probably the problem: no horizontal soft joints were installed in the building. In the photo, the steel angle supporting the face brick is lying directly on the brickwork below. The joint at the toe of the shelf angle is pointed with mortar and tooled. No space is left for vertical movement of the brick.

Without horizontal soft joints, the brick is forced to resist the shrinkage, creep, and elastic deformation the concrete frame under goes as it ages. As the concrete shrinks, the load on the frame is transferred to the masonry. Eventually, the load exceeds the strength of the brick and the brick cracks. If adequate soft joints had been installed under the steel angles during construction, the angles would have continued to transfer the weight of the masonry to the concrete frame instead of to the masonry below.

To solve the problem, replace the crushed brick and saw out $^3/_8$-inch joints under the steel angles. Fill the cut joints with backer rods and caulking, not with mortar. Install these soft joints at every floor.

In future buildings higher than three stories, be sure to install horizontal soft joints in brick veneer at each floor level. The shelf angle at each soft joint should not be continuous. Use 8- to 12-foot-long sections with a $^1/_4$- to $^1/_2$-inch gap between sections to allow for thermal movements of the angle. Also make sure the angles are properly anchored and shimmed so that they don't rotate and stress the masonry.

Installing rebar in block cores

Q. *Architects seem to be designing thinner walls with more reinforcing. We are having some trouble measuring the lap splice length of the rebars. Specifications require the lap splice to be 50 bar diameters or 12 inches, whichever is greater. When we puddle the bars into the grout we try to make sure the lap splices are the correct length. But we have no* *way of verifying this for the architect. Is there some nondestructive method for determining the lap splice length of vertical bars in grouted cores?*

A. The lap splice length of the rebars is important to the structural performance of a building. American Concrete Institute 531, Building Code Requirements for Concrete Masonry Structures, states that the minimum length of lap for bars embedded in grout, in tension or compression, shall be 0.002 d_b F_s, in inches, but not less than 12 inches. F_s is the full allowable stress in the reinforcement and d_b is the bar diameter. For Grade 40, F_s = 20,000 psi, and for Grade 60, F_s = 24,000 psi.

The lap splice length for Grade 40 bars, then, equals 0.002 x 20,000 x d_b or 40 bar diameters. For Grade 60 bars, the lap splice length equals 0.002 x 24,000 x d_b or 48 bar diameters. Regardless of the reinforcement grade, the splices must be a minimum of 12 inches.

Most specifications, including ACI 531.1, require that "reinforcement shall be supported and secured against displacement before grouting." While it is easier to grout the masonry units without the rebar in the wall, this practice violates most specifications. If you install the rebar after grouting, the architect may require you to demolish the wall and rebuild it, or he may ask you to verify that the lap splice lengths are adequate. Installing the bars after grouting also reduces the bond between the bar and the grout.

Verifying the lap splice length in an existing structure is difficult. Magnetic methods, such as a cover meter, have been used to successfully locate reinforcing bars in masonry structures. But they can't determine lap splice length. We are aware of two studies that were inconclusive.

In high-lift grouting, holes knocked into the masonry to clean out the cores before grouting also can be used to tie the rebar splice.

X-ray methods have been used to determine the position and condition of reinforcement, voids in concrete, and grouting of post-tensioned construction. Perhaps they also can be used to solve your problem.

The best solution to your problem may be to avoid it. Follow the specifications and install the vertical rebar before grouting. We can suggest three techniques to make this relatively easy. Two methods involve the installation and securing of the reinforcement. The third requires the purchase of special blocks. Check the building codes that govern in your area to make sure these methods are allowed.

The first method can be used in high-lift grouting operations. The procedure: Install the rebar after laying a one-story height of wall. Use the cleanout holes to securely tie the rebar splice, then pump the cores full with grout. The second method can be used in low-lift grouting operations where cleanouts are not used. The procedure: Lay up the first 4-foot height of wall. In the process, install bar chairs (various types are available) in cores where vertical rebar must be installed. Then insert 6-foot-long vertical rebars in the cores, properly positioning them in the rebar chairs. Next fill the cores with grout and repeat the process—lay another 4-foot height of wall.

This second procedure does not completely eliminate threading blocks over bars but it does minimize the problem. Note that the 6-foot bars provide a 2-foot lap splice. If a longer lap splice is required, longer bars must be used, making it more difficult to place block.

The third method can be used in low-lift or high-lift grouting operations. The procedure: Install and tie the vertical rebar first, then lay up the block. Use special A-shaped and H-shaped blocks. Made with one or two open ends, these blocks can be placed around the vertical steel instead of threading block over the top of the rebar.

In California, the State Division of Architects requires that the rebars be installed before the blocks are placed. This means special shaped blocks are the only acceptable, practical method. One note of caution: because of their shape, these special blocks are easily damaged during shipping.

Making joint reinforcement for corners

Q. *Do you know of a quick and easy way to make corner joint reinforcement from straight truss-type reinforcement?*

A. Prefabricated joint reinforcement is available for corners, T-intersections, and even circular walls. But if it costs too much or takes too long to get, you can make corner reinforcement from straight truss-type reinforcement in the following way:

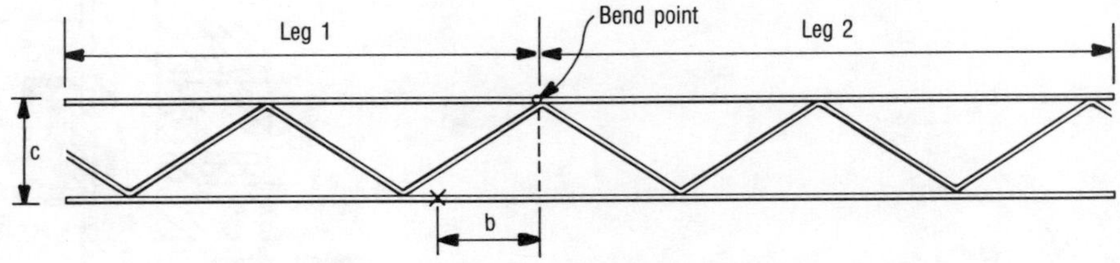

1. Cut wire here. The dimension b is equal to c, the distance between the longitudinal wires of the truss reinforcement.

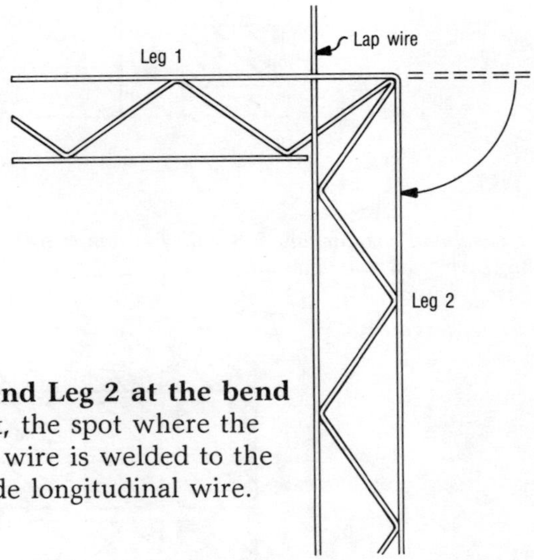

2. Bend Leg 2 at the bend point, the spot where the cross wire is welded to the outside longitudinal wire.

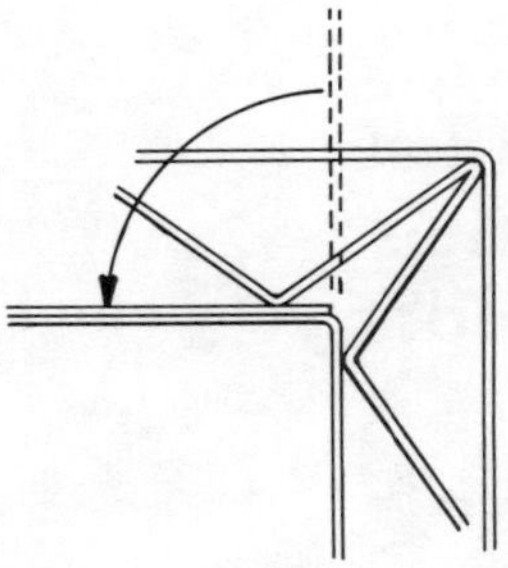

3. Bend the projecting cut wire so it laps the inside longitudinal wire of Leg 1.

Preventing grout blowouts

Q. *How can we prevent double-wythe walls from blowing out when we fill the cavity between wythes with grout?*

A. George Hanson, director of engineering for the Rocky Mountain Masonry Institute, says grout blowouts in double-wythe walls are most likely caused by:

- Not enough ties
- Ties not embedded enough
- Grout placed too fast in one location
- The vibrator left in one place too long

Space #9 wire ties no more than 24 inches longitudinally and no more than 16 inches vertically. For a double-wythe wall with a 2-inch cavity, embed the ties at least 2⁵/₈ inches *in each wythe.* In double-wythe walls, place grout in layers. Don't pump the grout in at one location and force it to run throughout the cavity. Move the vibrator continually; don't leave it in one place while you carry on a conversation or take a coffee break.

Don't project brick veneer too far

Q. *How far can brick veneer project beyond the toe of the shelf angle that supports it?*

A. In its Technical Note 28B, "Brick Veneer, Panel and Curtain Walls," the Brick Institute of America (BIA) says at least two-third of the brick thickness should bear on the shelf angle (see drawing).—*Courtesy Norbert V. Krogstad*

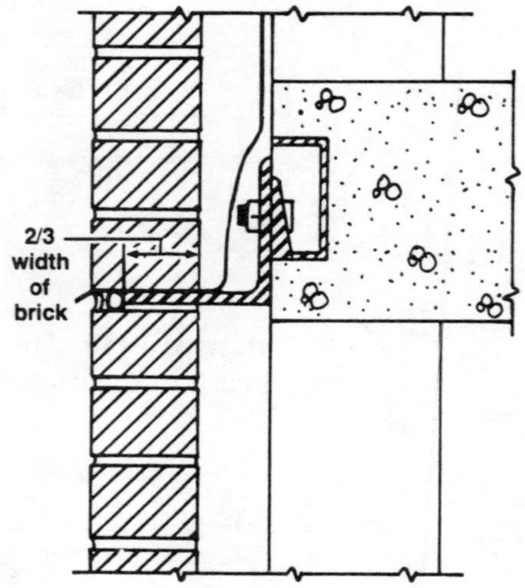

When must steel lintels be shored?

Q. *The specifications for my job call for temporary shoring of steel beam/plate lintels (see drawing). My last job did not require shoring for this type of lintel. Why do I need to provide shoring now?*

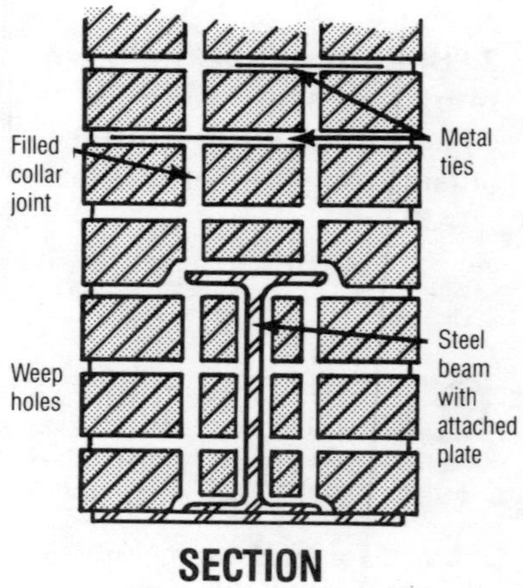

A. The structural design of the lintel determines whether or not shoring is needed. If the engineer's design considers the arching action of the masonry above the lintel, the lintel must be shored. Because arching action reduces the load on the lintel, considering it in the design allows the engineer to use a smaller (and less costly) steel lintel. But this smaller lintel must be shored until the masonry gains strength and the arching action occurs. The engineer specfies the shoring period based on the superimposed loads and curing conditions. Typically it is from 1 to 3 days.

If the engineer does not assume arching action, he must design the lintel for the full dead load and superimposed loads. The result is a larger steel lintel.

Because smaller lintels cost less, arching action is usually assumed and shoring is required. To reduce long-term deflections, however, some engineers even require large lintels to be shored. Because it is difficult to guess how the lintel was designed, always follow the specifications, even if they differ from job to job.

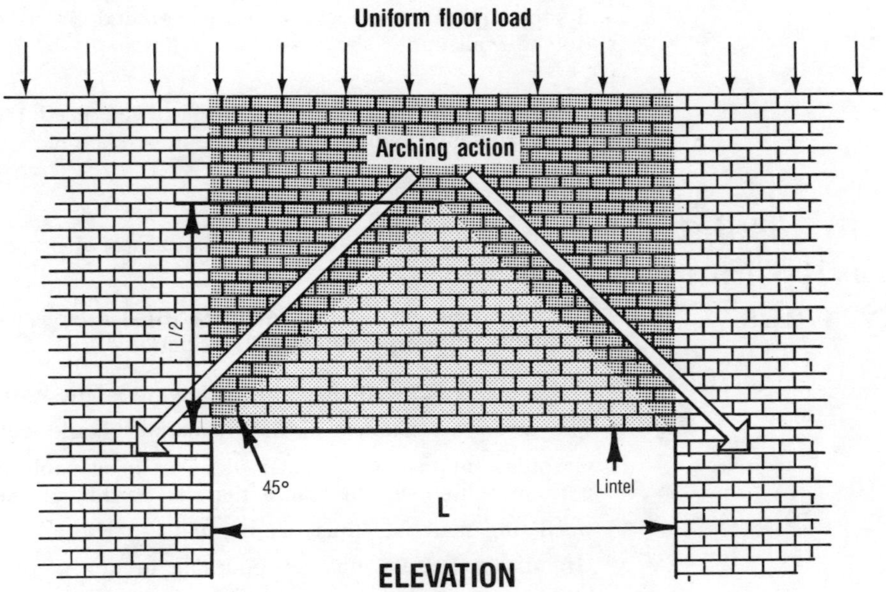

If arching action is considered in the design, only the load in the cross-hatched region, within the 45° lines, must be supported by the lintel. The superimposed floor load and the rest of the masonry load is transferred by arching action to the wall. If arching action is not considered, the lintel must be designed for the total masonry load and superimposed load, as shown by the shading.

Removing paint on soft brick

Q. *We've been asked to remove paint from the brick walls of an old building. We're a sandblasting company and we normally sandblast walls to remove paint. But sandblasting, we're told, could damage these soft brick walls. What other ways could we remove the paint?*

A. You are correct. Sandblasting soft brick would remove the surface of the brick as well as the paint. Removing the surface layer may create an unsuitable architectural appearance and also promote further deterioration. Future cleaning agents or air pollutants probably would penetrate the opened pores.

The method you should use depends on the type of paint and whether you're removing paint spots or paint from an entire wall. The paint type can easily be determined by sending a small flake of paint to a chemical lab. Once the paint is identified, an organic cleaner can be chosen.

If you're removing small paint spots, try using a poultice so you don't smear the paint over a larger area. A poultice is a chemical paste that dissolves and loosens the paint as the paste dries. The poultice also absorbs the paint, instead of letting it be reabsorbed by the masonry.

In general, only chemical paint removers, either alkaline strippers or organic solvents, can completely remove paint. Alkaline strippers are generally more economical and work particularly well on old linseed oil-based paints. They can cause efflorescence, however.

Organic solvents, such as methylene chloride, dissolve the organic components of the paint, causing the paint to swell and form a gel. Organic solvents tend to be more expensive, but they can dissolve most types of paint as well as urethane varnishes and epoxy. However, they also can spread spot stains deeper into the masonry and are toxic, requiring special handling and storing procedures.

Most manufacturers require that you wet the wall before you apply the chemical paint remover then rinse the wall thoroughly a specified time after applying it. Wetting the wall before minimizes the amount of cleaner absorbed by the brick. Rinsing afterward washes the chemicals away so they don't damage the brick. The chloride in organic solvents, for example, can attack the metal in the window frames or the reinforcing ties in the masonry. For this reason, all surrounding areas should be covered and protected from the chemical cleaner and the rinse water. Normally, you should rinse the walls until the wash water has a pH of 7, measured at the jobsite with pH papers.

To determine the best chemical cleaner, you need to contact a manufacturer of paint cleaning solvents and follow his recommendations.

How to insulate the cavity in cavity walls

Q. *What are the different ways (and costs) of insulating cavity wall spaces?*

A. The Brick Institute of America (BIA) established the following criteria for selecting suitable masonry cavity wall insulation.*

- The insulation must allow the cavity to drain water.
- The thermal resistance of the insulation should not be affected by moisture.
- Granular fill insulating materials must be capable of supporting their own weight without settlement, to assure that no portion of the wall is without insulation.
- Insulating materials must be inorganic: resistant to rot, fire, and vermin.
- Granular materials must be pourable in lifts of at least 4 feet for practical installation.

The table shows the types, thermal resistance values (R), and cost of common insulation materials. The costs are from Colorado insulation suppliers. Check with your local suppliers for the exact thermal resistance and cost of their products.

INSULATION MATERIALS FOR CAVITY WALLS

Material	Thermal Resistance, R (per inch)	Cost
Granular Fills:[1]		
Vermiculite	2.27	$4.25/bag
Perlite	2.70	$5.00/bag
Rigid Boards:[2]		
Expanded Polystyrene	5.00	$0.55/sf
Foil-faced Polyisocyanurate	7.20	$0.75/sf

[1] A bag contains 4 cubic feet.

[2] Price based on 2-inch-thick board.

*BIA Technical Note 21A, "Brick Masonry Cavity Walls Insulated," Brick Institute of America, 11490 Commerce Park Dr., Reston, VA 22091.

Random efflorescence

Q. *White efflorescence stains are appearing at random locations on a brick wall. I know efflorescence is salt drawn to the surface by water. But how do I determine the source of the salts and why are the salts deposited randomly?*

A. To determine the source of the salt, you should have the efflorescence salt deposits chemically analyzed by a testing laboratory. You also can do simple tests in the field. For example, if a drop of acid placed on the salt fizzles, it's probably calcium carbonate. Once you know what the salt is, you can determine its probable source using the table shown here from the Brick Institute of America (BIA). But you can't ever be totally certain. Water entering the wall also can carry salt with it.

Why is the efflorescence at random spots? It's probably **not** because the masonry contains random salt deposits. It's more likely the water drawing the salts to the surface is entering or exiting the masonry at random locations. This could be because at random spots mortar joints were partially filled, mortar beds were furrowed too deep, flashing or caulking was not properly installed, brick was picked up and relaid (thus breaking its bond), or the wall was not protected from weathering.

COMMON SOURCES OF EFFLORESCENCE

Principal Efflorescing Salt		Most Probable Source
Calcium sulfate	$CaSO_4 \bullet 2H_2O$	Brick
Sodium sulfate	$Na_2SO_4 \bullet 10H_2O$	Cement-brick reactions
Potassium sulfate	K_2SO_4	Cement-brick reactions
Calcium carbonate	$CaCO_3$	Mortar or concrete backing
Calcium hydroxide	$Ca(OH)_2$	Cement
Sodium carbonate	Na_2CO_3	Mortar
Potassium carbonate	K_2CO_3	Mortar
Potassium chloride	KCl	Acid cleaning
Sodium chloride	$NaCl$	Seawater
Vanadyl sulfate	$VOSO_4$	Brick
Vanadyl chloride	$VOCl_2$	Acid cleaning
Manganese oxide	Mn_3O_4	Brick
Iron oxide	Fe_2O_3 or $Fe(OH)_3$	Iron in wall

Source
"Efflorescence Prevention and Control, Part II," *Technical Notes on Brick Construction,* 23A, Brick Institute of America, 11490 Commerce Park Dr., Reston, VA 22091.

Design

3

Masonry

How high can concrete masonry screen walls be?

Q. *Plans for a new building specify a 4-inch-thick, 10-foot-high screen wall around an outdoor garden. The drawings show the concrete masonry wall is not to be reinforced. This seems a bit high to me for a nonreinforced wall. Are there standard design requirements for screen walls?*

A. You are correct. A 4-inch-thick, nonreinforced concrete masonry screen wall should not be built higher than 6 feet, 8 inches unless it is supported laterally in the horizontal or vertical direction every 6 feet, 8 inches. In the following table, the National Concrete Masonry Association (NCMA) gives the maximum spacing of lateral supports for different screen walls. The supports must be provided at these spacings in the vertical or horizontal direction, but not necessarily in both directions. Lateral support may be obtained from steel channels or beams or reinforced concrete masonry pilasters. If lateral supports are not provided, concrete masonry screen walls cannot be built higher than the distance allowed between lateral supports.

MINIMUM THICKNESS AND MAXIMUM SPAN FOR CONCRETE MASONRY SCREEN WALLS

Construction	Minimum Nominal Thickness	Nominal Thickness of Walls		
		4″	6″	8″
		Maximum Distance Between Lateral Supports		
Nonload-bearing				
Reinforced:*				
Exterior	4″	10′ 0″	15′ 0″	20′ 0″
Interior	4″	16′ 0″	24′ 0″	32′ 0″
Nonreinforced:				
Exterior	4″	6′ 8″	10′ 0″	13′ 4″
Interior	4″	12′ 0″	18′ 0″	24′ 0″
Load-bearing				
Reinforced*	6″	Not Recommended	12′ 6″	16′ 8″
Nonreinforced	6″		9′ 0″	12′ 0″

* Total steel area, including joint reinforcement, not less than 0.002 times the gross cross-sectional area of the wall, not more than two-thirds of which may be used in either vertical or horizontal direction.

Source: "Concrete Masonry Screen Walls," NCMA-TEK 5, National Concrete Masonry Association, 2302 Horsepen Rd., Herndon, VA 22070.

Building brick walls on wood floors

Q. *In remodeling my restaurant, I want to put an interior brick wall on the second floor. However, the local building officials say this cannot be done because it's against code to have a brick wall supported by wood framing. Is this true?*

A. It's not against all codes. Section 2515 (a) of the 1988 Uniform Building Code (UBC) says "wood members shall not be used to permanently support the dead load of any masonry or concrete." But exceptions are allowed. The same section of the code says that "brick veneer used as an interior wall finish may be supported on wood floors which are designed to support the additional load, and be designed to limit the deflection and shrinkage to $^1/_{500}$ of the span of the supporting member." The building code in your area may be different—or perhaps the building officials overlooked this exception. If your local building code does allow brick veneer walls to be supported by wood floors, check with a structural engineer to make sure your wood floor can support the wall you want. If it doesn't, he may be able to recommend modifications to the floor or wall that would satisfy code requirements.

Considering corrosion in steel lintel design

Q. *As an engineer, how do I design steel lintels for exterior brick masonry so that corrosion is not a problem?*

A. To compensate for corrosion, many engineers specify steel lintels with a plate thickness of at least $^1/_4$ inch or even $^5/_{16}$ inch. If the steel area required by design is greater than $^5/_{16}$ inch, some engineers add an extra $^1/_{16}$ inch to allow for corrosion. According to Fontana and Greene, *Corrosion Engineering,** typical corrosion rates for steel exposed to the atmosphere, industrial and marine, are 1 and 5 mils per year, respectively. A mil is 0.001 inch thick. An extra $^1/_{16}$ inch (62.5 mils) of steel adds $62^1/_2$ years of service in industrial environments and $12^1/_2$ years of service in marine environments.

The Brick Institute of America (BIA) suggests using galvanized steel lintels in harsh climates, such as in marine exposures.† Galvanized steel lintels do not require the periodic maintenance often required by ungalvanized steel lintels. Galvanized steel lintels cost more, however. In Colorado, they can cost about 28¢ per pound more than plain steel lintels.

Because harsh climates are never well defined, we suggest you inspect steel lintels on some old buildings in your area. If you find little corrosion, specify ungalvanized steel lintels. Select the lintel that meets the minimum steel area required by design—or to be safe increase the thickness by $^1/_{16}$ inch. If you find extensive corrosion, specify galvanized lintels.

---

* Fontana and Greene, *Corrosion Engineering,* 2nd Ed., 1978, McGraw-Hill, Inc., New York.

† BIA Technical Note 31B, "Structural Steel Lintels," Brick Institute of America, 11490 Commerce Park Dr., Reston, VA 22091.

Expansion joints needed

Q. *I live in a subdivision that is surrounded by a brick fence. Enclosed is a photo of the brick wall. You can see that there is a lot of cracking. Many subdivisions have the same type of wall, but they aren't cracked as bad as mine. Why did the wall crack?*

A. More information on the design, materials, and construction of the wall is needed to determine the exact cause. However, it's likely that movement caused by temperature changes has caused the damage.

Like all materials, masonry expands as the temperature increases and contracts as the temperature decreases. Depending on the type of masonry and the range of temperature it's exposed to, masonry walls usually should be built with expansion joints located at least every 100 feet. Your photos don't show any expansion joints.

The configuration of the wall might also have contributed to the cracking. During thermal expansion or contraction, the joint between offset walls is highly stressed. If there is no joint at this location or the joint is not properly installed, the wall cracks to relieve this stress.

Brick paving: Use mortar or not?

Q. *In exterior applications, is it better to use mortar between brick pavers or to place pavers without mortar?*

A. No simple answer to this question applies to all uses of pavers in all parts of the country. In general, brick paving without mortar is more appropriate for northern climates, where mortar can deteriorate rapidly due to freeze-thaw action. Brick paving without mortar and with a good subsurface drainage system holds less water and performs better. Mortarless paving systems have other advantages, which also are factors in southern climates. Mortarless brick pavements accommodate thermal and moisture expansion between individual pavers, reducing the need for expansion joints. They also accommodate small amounts of differential settlement without cracking. Mortarless brick pavements have some disadvantages, however. Dirt that collects in joints between pavers can support plant growth. Also, if the exterior paving is an extension of an interior brick floor, it may not be architecturally appropriate to change from a mortared system inside to a mortarless system outside.—*Courtesy Norbert V. Krogstad*

Specifying bond strength

Q. *I would like to require a minimum ultimate bond strength in my project specifications. What value for bond strength should I require?*

A. No standard now sets minimum values for ultimate bond strength of masonry. Building codes list only allowable values for bond for use in design. Until some standard sets minimum values (ASTM is now attempting to do that), I suggest requiring a minimum 90 psi for Type S and 70 psi for Type N portland cement-lime mortars, tested according to ASTM C 1072, Standard Method for Measurement of Masonry Flexural Bond Strength. You can require somewhat higher values if higher bond strength is needed for structural reasons. Lower values might adversely affect wall permeability, mortar durability, or other important performance characteristics.

I base these recommendations for minimum ultimate bond strength on an unpublished 1964 study by the Structural Clay Products Institute (now the Brick Institute of America). In the study, 135 test were conducted using Type S portland cement-lime mortar and 27 different types of brick with initial rates of absorption (IRA) from 1 to 50. All samples were cured for 28 days in laboratory air. The average bond strength in the study was 123 psi with a standard deviation of 32 psi. Assuming a normal distribution, 85% of the tests would be greater than 90 psi. A similar procedure was used to determine a minimum value for Type N mortar. Though the 1964 study used what later became ASTM E 518, Test Method for Flexural Bond Strength of Masonry, C 1072 will give higher results than E 518.—*Courtesy Norbert V. Krogstad*

Eliminating odd size bricks around openings

Q. *In a building we're designing, a 6-foot-wide door opening must be centered in an 11-foot wall. This leaves a $2^1/_2$-foot wall on each side of the opening. To lay this wall using modular 8-inch-long brick in a running bond, the masons have to cut 2- and 6-inch-long bricks to fit around the opening. Is there any way to avoid all these odd size pieces and the effect all these cut units have on the appearance?*

A. Heights and lengths of walls and distances between windows, doors, piers, control joints, and other interruptions in the wall plane should be based on the dimension of the masonry unit that is used. When an opening or other building element cannot be relocated, try compensating by using a special size unit in stack bond. For the 6-foot-wide doorway you describe, consider using a 6-inch-long accent brick, decorative block, or stone in stack bond around the opening (see drawings next page). This would solve the bonding problem and enhance the design at the same time. Examples of this are the brick detailing around windows of the Stearns County/St. Cloud Law Enforcement Center, St. Cloud, Minnesota, and the concrete block detailing around the windows of the Mechanics Bank in Sturbridge, Massachusetts.

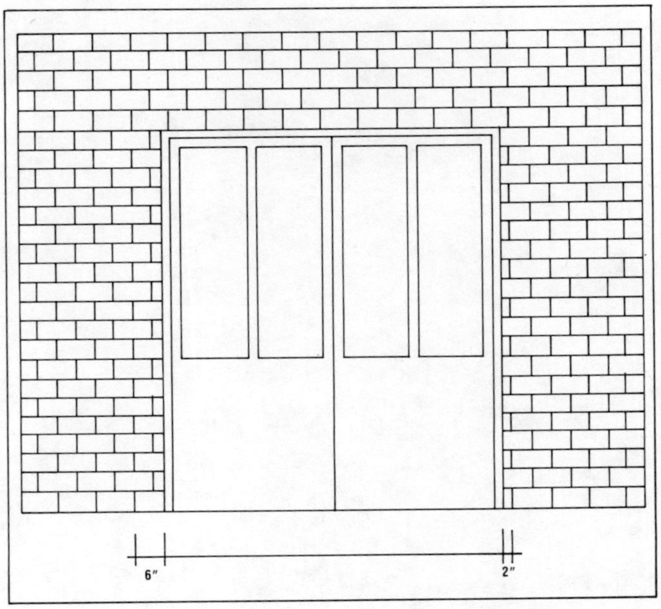

Special size units in stack bond (top) eliminate cutting odd size units (bottom) around openings that cannot be relocated to fit the modular dimension of the brick.

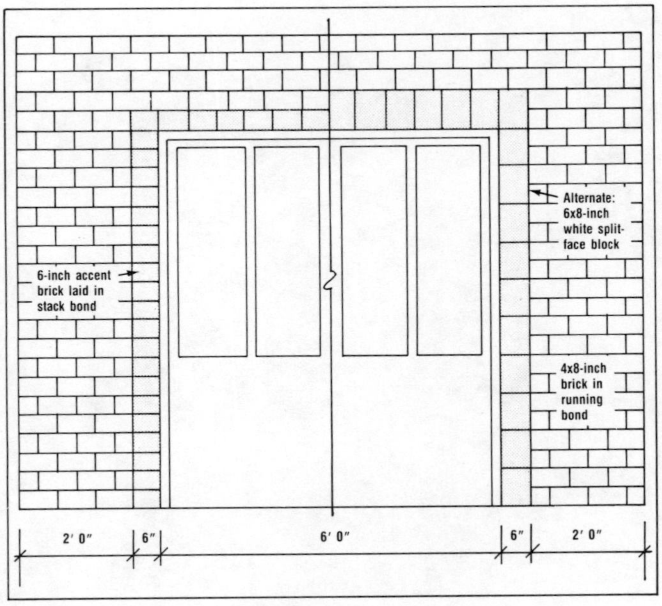

Index

Masonry